Torbern Olof Bergman

Thoughts on a Natural System of Fossils

Part I

Torbern Olof Bergman

Thoughts on a Natural System of Fossils
Part I

ISBN/EAN: 9783337022457

Printed in Europe, USA, Canada, Australia, Japan

Cover: Foto ©berggeist007 / pixelio.de

More available books at **www.hansebooks.com**

THOUGHTS

ON A

NATURAL SYSTEM

OF

F O S S I L S.

Res ardua, vetuftis novitatem dare, novis auctoritatem——dubiis fidem, omnibus vero naturam et naturæ fuæ omnia.

PLINIUS.

PART I.

ARRANGEMENT OF FOSSILS.

NATURAL BODIES IN GENERAL.

§ 1. *Principal Divifion of Natural Bodies.*

ALL bodies which nature fpontaneoufly produces upon the furface of the earth may be properly divided into *organifed* and *unorganifed.*

§ II. *Organifed*

§ II. *Organifed Bodies.*

THESE are poffeffed of a number of internal veffels, by which, from the nourifhment they take in, the particles neceffary to the increafe, fupport, and propagation of fuch bodies, are extracted, prepared, conveyed, and diftributed.

§ III. *Claffes of organifed Bodies.*

THESE bodies are diftinguifhed by the epithet *living ;* and, whether they poffefs fenfibility or not, they conftitute two immenfe claffes, the *animal* and the *vegetable,* which are commonly confidered as two diftinct kingdoms in nature.

§ IV. *Unorganifed Bodies.*

THOSE bodies are termed unorganifed that are entirely without any organic ftructure, and feem to be formed by the accumulation of particles united folely by the external force of attraction.

§ V. *Various*

§ v. *Various Confiftencies of unorganifed Bodies.*

THESE differ in many refpects, but we fhall here take notice of the degrees of *denfity* only, which has commonly been defigned by the name of confiftence.

§ VI. *Solid Bodies.*

CONSIDERING thefe, **then,** according to this rule, we find fome bodies fo *folid*, that their particles are fo firmly united as not to be feparated but by a very confiderable force. Of this kind are moft of the foffils.

§ VII. *Liquid Bodies.*

SOME again are *liquid*, whofe component parts adhere fo loofely, that they may be feparated by the fmalleft impulfe ; but being left undifturbed, they, by the force of gravity, arrange themfelves in fuch mutual equilibrium, as to prefent a furface parallel always to the horizon.

§ VIII. *Fluid*

§ VIII. *Fluid Bodies.*

OTHER bodies are reckoned *fluid*, whofe particles are not only eafily feparable, but feem in fome degree to repel each other. It is true, they feek an equilibrium; but, as they are not lefs influenced by elafticity than by gravity, they oftener appear with the unequal furfaces we daily fee in clouds and vapours.

§ IX. *The Utility of this Diftinction.*

ALTHOUGH the fame body, as occafion requires, may undergo every variation of confiftence, yet this diftinction is not the lefs to be regarded; for peculiar qualities, with a confiderable difference in their proportions belong to each condition. But the plan we have propofed to follow, will not admit of a further explication of this matter.

§ X. *The continued Series of Natural Bodies.*

THE great Leibnitz, by that law to which he gave the name of *continuity*, denied formerly that there could poffibly be any interruption be-

tween

tween phyfical **caufes and** effects; and maintained, with fuch confidence, its invariable ope-
ration and influence, that he predicted, that
fome time or other a fpecies of animals (as the
zoophyta) would be difcovered, partaking more
or lefs of the nature of vegetables.—The celebrated Trembleyus, by the difcovery of the Poly-
pi, afterwards confirmed the truth of this pre-
fage. Daily experience alfo convinces us of the
exiftence of fuch a connecting chain in the order of natural bodies; fo that, though we are
acquainted with feveral links fingly, yet it may
feem fcarce poffible to afcertain thofe that fhould
be immediately united to them.

§ XI. *The Neceffity of a Syftem in* **Natural**
Hiftory.

As natural bodies **may in** various ways be
rendered ufeful to man, a thorough knowledge
of them becomes highly neceffary; and it will,
indeed, in general be found, that their utility
encreafes in proportion **to** the extent of that
knowledge. Their great number and variety
require fyftematic arrangement; without which
the neceffary diftinctions could not be made,
and which, in fome cafes, where the difference
is very minute, would be productive of great inconvenience.

O

§ XII. *Criteria*

§ XII. *Criteria of Natural Bodies.*

In order to difcriminate with fafety and precifion, even where bodies are united in the greateft affinity, it is an object of the firft importance
to eftablifh proper *criteria*.

§ XIII. *Conftant and perpetual Forms* **of Organic**
Bodies.

In the egg, or in a fecundated germ, the
little body, the rudiment of the future fœtus,
lies wholly concealed, until by proper heat and
nourifhment it is gradually evolved, increafes,
and arrives at maturity. In all organic bodies,
therefore, the form is predetermined from their
very origin, which the power of their internal and
peculiar ftructure is calculated to develope ; fo
that between thefe two qualities the relation is
invariable ; and therefore criteria are not impro-
perly collected from that external figure which
is derived from, and rooted in the effential character of the fpecies.

§ XIV. *Monftrous Productions.*

Among thefe, indeed, we fometimes find de-
viations from the general laws of nature, producing *monfters* ; but fuch events which are rare,
and

and arising from particular causes, are almost always unlike each other.

§ xv. *Fossils.*

ALL unorganic bodies, as well solid as liquid, which are either altogether without any organic structure, or display the ruins only of organization, are denominated *fossils*, or more commonly *minerals*.

§ xvi. *The Mineral Kingdom.*

THE term *fossil*, or *mineral kingdom*, is generally applied to an arrangement of such fossils as are found in the earth.

§ xvii. *Generation of Fossils.*

IN this third kingdom of nature, the process of generation is carried on in a manner widely different from that of organized bodies. Here is no egg, no feed, to cherish and support the future fossil, confined and restrained within the narrowest limits; no fecundation; no established circulation of the nourishing fluids; nor any evolution. Molecules uniting, by the sole

O 2   power

power of attraction, form at once the growth
and perfection of foffils.

§ XVIII. *Variable and inconftant Form* **of Foffils.**

FORM, and other external qualities, of **which**
the fenfes only can determine, depend upon
circumftances that are perpetually varying, but
which do not in the leaft affect the intrinfic nature of the foffil.

The pofition may poffibly need the illuftration of an example. Let us take **a quantity** of
water, charged with aerated calcareous particles, and we fhall fee arife various figures, tex-
tures, and cohefions, according to the different
modes in which the concretion was performed.
By the fubfidence only of the atoms a cruft is
generated, parallel to the bottom, if the diftri-
bution of them has been **made** equally through-
out the whole mafs; if otherwife, the greater
part forms tubercles farther from the furface of
the bottom, **than** in the fuppofition of equality.
Water impregnated with aerial acid acts like a
menftruum; and, though it does not at all affect
the faturated particles in this hypothefis, yet it
neverthelefs has confiderable influence in forming their concretions. Such water oozing
through fubterraneous **vaults,** generates calcareous drops, hanging from the roof, while
pointed

pointed cones are produced by the falling fluid
upon the floor, and both increafing in length,
meet at laft, and form one continued column.—
If the fame water purfues its trickling courfe
along the walls, we find them covered with a
ftalagmitic cruft; which according to the diverfity of the protuberances exhibits a great variety of figures, that, with the affiftance of a warm
imagination, may be made to refemble complete animals, or their feveral members, and a
thoufand other forms and appearances.—From
this water fuffered to remain long at reft, fpataceous cryftals are feparated, that affume various fhapes; as the granatic, the fchoerlaceous,
hyacinthic, dodecaedric, and thofe pyramidal on
both fides, named fwines teeth,—and many o-
thers.

The internal texture likewife admits of confiderable variation. The moft fubtle particles
unite into a denfe and equal mafs: Thofe that
are granulous, and of many angles, form combinations more rough and uneven; fuch as are
produced by chryftallization appear fpataceous;
and others that are alternately depofited in ftrata, or lamellæ, prefent a divided ftructure.

The degrees of cohefion alfo vary according
to circumftances. Water charged with fine
particles of aerated chalk, and quickly evapora-
ted, leaves a powder fcarcely cohering, and
which foils the fingers, like the mineral known

O 3

by

by the name of Agaric. Larger maffes however of calcareous powder, expofed for many years to the preffure of a confiderable weight, acquire at length fuch a degree of confiflence, that diftinct lines can be drawn with fmall pieces of them; indeed this property is found in calcareous chalk likewife.—Hitherto the greater degree of hardnefs has been produced by cryftallization, as we find that calcareous cryftals make no mark whatever, a circumftance owing to the firm union of their particles, by which the friction on a painter's canvafs has no effect upon them, at leaft fo as to be vifible.

What has been thus briefly ftated may be fufticient to fatisfy us, that, from the external qualities of foffils, no proper judgement can be formed of their internal compofition.

OF THE SEVERAL CRITERIA OF FOSSILS.

§ XIX. *Oryctology.*

ORYCTOLOGY, or Mineralogy, are names given to that fcience, which fo arranges all the known foffils, that they may be accurately diftinguifhed from each other.

§ XX. *Various*

§ xx. *Various Systems of Oryctology.*

As zoologists, in their arrangement of animals, have chosen different parts; some the feet, others the teeth, the becks, and other parts, according to the agreement or disagreement of which their different systems were established; and, as botanists have differed in the principles of their science, one preferring a leaf, another the petals, a third the stamina and pestillum, while a fourth maintains the superiority of the fruit;—even so is it with mineralogists, who have often pursued very different paths, in their endeavour to illustrate and confirm the same object. Such a view of natural bodies, taken as it were from many different points, has however its advantages, as it increases the number of accurate comparisons. But, as every method cannot equally answer the end proposed, it becomes necessary to select that which is the most perfect and convenient.

§ xxi. *The best Arrangement.*

As, in order to understand the nature of fossils, and apply them to purposes of utility, it is necessary to arrange them in some kind of systematic order, the preference is certainly due to that method, by which both their internal cha-

racter

racter and compofition may be made equally e-
vident. Effential properties depend on the quality of the parts that enter into compofition,
and their mutual proportion; and, unlefs we
are well aquainted with thefe parts, we fhall labour to little purpofe, in our attempts to mould
them to our own defires: Nay, we often meet
with difappointments, becaufe we have not confidered that our views are inconfiftent with the
very nature of the materials fubjected to experiments.

§ XXII. *In what manner the Compofition of Foffils may be afcertained.*

HAVING fettled thefe points, it remains yet to
be determined in what manner we are to judge
of the *compofition* of foffils: Whether the connexion between fuperficial marks, and the intrinfic character, is fo intimate and confequent, that
the former cannot be known, without the other
being revealed? whether it may be neceffary
to proceed **by a** chemical analyfis in the dry
way? or, fhould this not be fufficient, are **we to**
have recourfe to the moift way? We will **con-**
fider thefe queftions feparately.

§ XXIII. *External Criteria.*

IF, through the means of criteria collected
from

from the external appearance, and obvious to all, we were able to obtain the object of our research, no method could certainly be more simple; for, with the affiftance of our fenfes only, we might difpenfe with the tedious procefies of experiments: But we have already difcovered the fallacy of relying on many of thefe marks, even the moft principal, as they are liable to be affected by various circumftances of fituation, and diverfified without end, (§ xviii.). It may be proper, therefore, to enter a little more minutely into the confideration of this queftion.

§ xxiv. *Uncertain and deceitful Size of Foffils.*

In no criteria can we poffibly have lefs faith than in that of magnitude; and we cannot fufficiently exprefs our aftonifhment at the violence offered to nature, when a larger piece of ftone, referred to its proper genus, if reduced to a powde, is not only exiled to fome other, but is not even permitted to remain under the fame clafs.

§ xxv. *And Colour.*

The vulgar proverb, that cautions us againft belief in colour, is not inapplicable to oryctology. It is well known, that there are feven primitive colours; and, in order that a body appear coloured, it is requifite that fome particular

cular kinds of rays be reflected; would we en-
quire into the caufe of this phenomenon, we
muft feek it in the quality of the furface, which
is indeed often fo tranfient, that the colour may
be changed, or entirely deftroyed by the heat
of boiling water, or even by the influence of fo-
lar light.

A tranfparent colour arifes from tranfmitted
rays, and feems to indicate a fpecies of attraction; while, on the other hand, an opaque colour implies repulfion. Both without doubt fug-
geft the idea of fome relation between the light
and the given body; but which is of fuch fubtlety, that though it alone were varied, the character of the matter remains altogether unalter-
ed; at leaft the difference is not obvious to the
fenfes. We have feen, that tranfparency de-
pends upon the difpofition of the particles; and
this once difturbed, the tranfparency vanifhes,
and with it all the effect produced by tranfmitted rays. Thefe feveral appearances feem to a-
rife from the phlogiftic molecules, which vary either as to quantity, magnitude, or elafticity. Ve-
locity even determines the difference of colours.

§ XXVI. *Internal Texture* **and Form.**

WE have already touched on *internal texture
and form* in the foregoing divifions, (§ xviii.)
Determinate

Determinate figures bear a refemblance to geometric bodies, and it is not without fome degree
of probability that they are faid to be derived
from the nature of the matter : An opinion that
has long influenced many to believe, that certain figures were proper and effential to different fubftances. The folly of this doctrine I
have elfewhere demonftrated at large *. If
therefore regular figures, and thofe beft defined,
are fallacious, we are furely not to rely on any
fuperficial characters which are very often common to fubftances of the moft oppofite qualities,
and never uniformly conftant in the fame fpecies.

§ xxvii. *Phyfical Marks of Earths.*

Nor are we wholly to neglect the *phyfical
marks*, which, though they cannot be fully efti-
mated by the external fenfes alone, yet may be
afcertained by eafy experiments, without the
trouble of decompofition. Such, in the firft
place, are hardnefs and fpecific gravity; to
when, indeed, we may add the relation to the
magnet.

§ xxviii. *Hardnefs.*

Degrees of hardnefs may be determined in
various ways, by the nail, the knife, or by fteel;

and,

* Effays, **vol. 2.**

and when they are more intenfe, by a feries of gems, cut exprefsly for this purpofe. But this property indicates lefs the matter, and its mixture, than the various exficcations arifing from different circumftances, the fubtlety and cohefion of particles, denfity, and fuch like. Soft clay dried gradually, and afterwards expofed to an encreafing fire for feveral hours, until it is brought to a white heat, becomes **harder and** harder, and is at length capable, like a flint of ftriking fparks from fteel. In all this procefs, however, the matter is no otherwife affected **than** by a contraction of its bulk, which is diminifhed about **one half.**

§ xxix. *Specific Gravity.*

SPECIFIC gravity is determined by the hydroftatic balance, which properly indicates nothing elfe than the denfity or quantity of matter in a given volume. A knowledge of this property is of confiderable utility, efpecially in the examination of metals, whether pure, or of known mixture ; but with refpect to other foffils, **the** difference is fo very triffling, that their nature and compofition can fcarcely ever be this **way** afcertained.

§ xxx.

§ xxx. *Examination by the Magnet.*

IRON, unlefs it is dephlogifticated below a
certain point, is ever obedient to the magnet;
but this mark is particular. Various phenomena likewife authorife a fufpicion that many o-
ther fubftances are attracted by it; therefore no
reliance can be had upon this as a diftinguifhing
character.

§ xxxi. *Real Utility of external and phyfical*
Marks.

ALTHOUGH fuperficial criteria contribute nothing to the true knowledge of foffils, and that
the obfervation of Juvenal, *fronti nulla fides*, may
be well applied to them, even though the phyfical properties be at the fame time underftood,
(§ xxviii. xxx.) yet we are not altogether to pafs
them over in contempt. By fuch accurate de-
terminations as the celebrated Werner fo fuccefsfully attempted, they are rendered very proper for diftinguifhing varieties; and when the
eye is once habituated to them, they often lead
it directly to diacritic experiments. Perhaps the
compofition being thoroughly afcertained by a-
nalyfis, an exact comparifon may affift confiderably in drawing a juft inference.

§ XXXII.

§ XXXII. *Nature of Foſſils diſcoverable by the Aid of Chemiſtry.*

In order to diſcover the proximate principles of foſſils, it is neceſſary to have recourſe to chemical experiments. But will not the ſimpler kinds be ſufficient, in which the foſſils, whether alone, or with the addition of proper fluxes, are melted in the fire and treated in various ways? This indeed is the path purſued with indefatigable zeal by the celebrated Pott, and which no one ſince him has extended **with** more ſucceſs **than the renowned Monſieur D'Arcet. How far it is connected with our deſign we ſhall preſehtly have occaſion to obſerve.**

§ XXXIII. *Their Character in the Fire.*

A THOROUGH knowledge of the effects produced by fire upon foſſils is of the greateſt importance in the cultivation of many arts. For if we recollect that bricks, tiles, crucibles, glaſs, **a**maufa, earthen and china veſſels, eliquation **of** metals, and other works, can neither be carried on nor completed without the aſſiſtance of fire, we ſhall ſee that this knowledge **is** equally neceſſary and extenſive.

§ XXXIV.

§ XXXIV. *Use of the Blow-Pipe in Oryctology.*

Nor can we pass over in silence the great u-tility of the blow-pipe in oryctology, by its speedy and concise mode of operating. With it a few minutes are sufficient to examine the nature of a fossil, upon a piece of coal, or in a spoon of gold, and to observe all the changes from beginning to end; which for the most part is not possible in a crucible; notwithstanding in this way, it requires several hours before the result of the process can be known *.

§ XXXV. *Most of the Principles of* **Fossils are** *discovered by Fire.*

It must, however, be acknowledged, **that,** in many cases, the principles of fossils may be ascertained by the proper application of fire; unless, by the number or delicacy of such principles, the composition of the fossil is rendered too complex and intricate.

§ XXXVI. *But not every Principle.*

There are many circumstances that will prevent us from considering fire as the supreme arbiter of composition, though supported with all the

* Essays, 2d **vol. page 455.**

the affiftance of the dry way; and it may be fufficient to enumerate fome of the moft confiderable.

§ xxxvii. *Why Inveftigation by Fire is fometimes fallacious.*

FIRE tends to confound all principles together, except thofe of metallic bodies which are feparated from their matrices; it is therefore **not** at all calculated to extricate the feveral ingredients of compofition.

§ **xxxviii.** *The Efficacy of Fire cannot be defined with any certainty.*

AN accurate and eafy meafure of the power of this element is yet wanting. A foffil refifts a certain degree of heat, that will yield to one more intenfe; and there are perhaps a very few that are deemed altogether refractory.

§ xxxix. *And it is variable alfo.*

IT is **not** uncommon for the **fame** degree of fire to **melt** fome varieties of the fame **fpecies**, while upon others, it feems **not** **to have the** fmalleft influence. The **petrofilices, feltfpat,** and other foffils, **afford examples of this** kind.

§ XL.

§ XL. *Does not determine the Proportion of the different Principles.*

AND laftly, if fometimes it is competent to dif-cover fingle principles, yet it always **conceals** their mutual proportions. This imperfection is of the greater moment, as it is evident, that the proportions of the fame materials being varied, both the appearence in the fire, and the other qua-lities of the foffil, are often confiderably altered.

§ **XLI.** *Merit of Cronftedt.*

THE celebrated Cronftedt, in his **excellent fy-**ftem of foffils, has eftablifhed the fuperiority of principles, and has therefore conceived the ge-nuine method ; and if, notwithftanding, he has occafionally fallen **into** errors, they muft be at-tributed to the **want of** proper experiments.

§ **XLII.** *The beft Method of examining Foffils in the Humid Way.*

THE illuftrious Margraf had no fooner difco-vered the true method of decompofition, the hu-**mid** and menftrual, than he endeavoured, by his own exertions to render it eafy and practicable. The **new** road into which he ftruck, was befet with thorns and briars ; but it is certainly the

P   only

only one that leads to a knowledge of principles, both as to quality and quantity; and therefore indifpenfably neceffary in every enquiry into compofition.

§ XLIII. *The Difficulty of founding* **a Syftem of**
Foffils.

IT was the opinion of the celebrated Lehman,
whofe judgement in fuch matters was unquefti-
onable, that a thoufand years would not be fufficient for the conftruction of a fyftem of foffils,
arranged according to proximate principles, on
account of the immenfe number of various foffils,
and the daily augmentation it is receiving; the
variety and expence of the neceffary experiments, and the want of a more general fpirit
of adventure and induftry requifite for fuch an
undertaking.

§ **XLIV.** *Internal and external Characters.*

A collection of thofe properties on which the
leading principles depend, is called the *internal
character;* and the chief fuperficial marks of a-
ny foffil taken together, conftitute the *external
character.*

Or

Of the Classes of Fossils.

§ XLV. *Enumeration of the Claſſes.*

AVICENNA, an Arabian phyſician of the e-
leventh century, divided foſſils into the four
claſſes, of ſalts, earths, metals, and phlogiſtic
bodies. In this diviſion, all ſubſtances agreeing
either in external or internal character, are properly enough combined ; and, as hitherto no
general arrangement has been propoſed preferable to this, it is no doubt worthy of being
continued.

§ XLVI. *Order.*

THE order of the claſſes may in a great mea-
ſure be treated as a matter of indiſference ; however, I think it right to begin with Salts, as being the only ſubſtances ſoluble in water, and
which ought to be thoroughly underſtood, in
order to develope the nature of the other claſſes;
and perhaps, becauſe they are radically united
with each of them, though the moſt conſidera-
ble number of them have as yet in this ſtate e-
ſcaped diſcovery.

Phlogiſtic bodies I place the laſt in order ; for
theſe by their prevailing principle approach nearer than any of the other claſſes to organiſed bo-

dies,

dies, charged with inflammability, and to which
principle foffils perhaps are indebted for their
exiftence. Earths and metals, according to their
character, hold **with** propriety a middle fta-
tion.

§ **XLVII.** *Diftinguifhing Marks of each* **Clafs.**

For the prefent it may be fufficient to **men-**
tion the following criteria of the claffes, **which**
fhall afterwards be more fully explained.

Salts very finely pulverifed, and diffolved in
a thoufand times their weight of water, are more
or lefs fenfible **to the** tafte. With refpect to
diftilled water 2 **is the** common limit of their
fpecific gravity.

Earths have neither tafte nor folubility. They
are however taken up by proper fimple falts.
Though for the moft part heavier than falts,
they are not reducible to a metallic ftate. When
compared with water, their fpecific gravity
fluctuates between 3 and 4½, which it has never
yet exceeded.

Metals are not foluble in water; have a peculiar fplendour; and furpafs all other known bodies in fpecific gravity. They are at leaft fix
times heavier than equal bulks of water, commonly much more; but never exceeding twenty times.

Phlogiftic bodies are almoft always lighter

than

than the falts ; but have this peculiar quality of being combuſtible.

§ xlvⅢ. *Taſte.*

TASTE, depending upon the fenfibility of the tongue, differs fo much in different perſons, that what will excite powerful fenfations in one man fhall not be at all perceptible to another. It is evident, therefore, we are to place but little dependance on this quality.

§ XLIX. *Solubility in Water.*

SOLUBILITY in water, confidered generally, is an *unlimited property.* In order to define it, it will be neceſſary to attend to the ſtate of diviſion of the body to be diſſolved, and the quantity and temperature of the menſtruum employed.

Pulverization encreaſes the extent of ſurface ; and in proportion as it does fo, the menſtruum, by coming into contact in a greater number of points, acts with more efficacy. For this reaſon large maſſes immerſed in a menſtruum, are fometimes very little, if at all corroded : When divided into fmall pieces they offer leſs reſiſtance; and, if pulverized are entirely diſſolved. It happens occaſionally, however, that mechanical diviſion does not anſwer the end effectually,

and

and therefore recourfe is had to the more fubtile powers of chemiftry; and the precipitation of a folution made in a ftronger menftruum, is taken fuccefsfully for this purpofe. For a precipitate yet moift and recent is fo open and fpongy, that it far exceeds all mechanical divifion.

In like manner, though a folution cannot be effected in an equal weight of water; yet, if that weight is doubled or tripled, or fufficiently encreafed, there would be no doubt of producing it.—If water of a moderate temperature avail nothing, tepid or warmer water may fucceed; and fhould this degree alfo of heat be ineffectual, it may yet be raifed to fuch a height in a clofe veffel, as will generally overcome all refiftence, and even produce effects fcarce to be expected.

Hence, then, I apprehend it is evident, that the very nature of folubility will not admit of any certain or determinate criteria, but that it may be faid rather to proceed in an infinite feries: For if, on inftituting an experiment, nothing is diffolved, a fufpicion will always arife that if the refifting matter were either more minutely divided or immerfed in a greater quantity of water, or in water of a higher temperature, it would neceffarily be diffolved. In this manner, therefore, all certainty is deftroyed, and every conclufion rendered merely conjectural.

§ 4.

§ L. *Artificial Limits of Solubility.*

IF folubility ever becomes an ufeful criterion, it muft be by affigning to it certain neceffary artificial limits. Having duly confidered this idea, I have pronounced thofe to be the beft, that can be found moft eafily every where. I have felected therefore for this purpofe mechanical pulverization, a weight of water a thoufand times heavier than the fubftance to be diffolved, and a degree of heat equal to boiling, as boundaries more proper than any others.

§ LI. *Great Extent of Solubility.*

WE are very far from believing that **this li-**mit is to interrupt one link in the great connecting chain of nature. Our ignorance and weaknefs have rendered it neceffary ; and, whatever fubftances beyond it a more improved ftate of fcience may difcover, we fhall refer them to the clafs of earths, though we give them the appellation of faline, as an indication of their character. As examples of fuch faline fubftances, we may take the filiceous earth, which is found abfolutely diffolved at Geyfer in Iceland * ; and the zeolithic, at Laugarnaes in the fame ifland †.

P 4   Vitriolated

* Effays, vol. 3d. p. 251.
† Ibid. p. 255.

Vitriolated ponderous earth, commonly called *fpatum ponderofum*, aerated lime *, fluorated lime, impregnated with the acid of the *lapis ponderofus* †, are all faline earths, by the force of compofition, and are even without doubt foluble, though to what extent experience has not yet determined.

§ LII. *Diftinguifhing Marks of Earths.*

THE characters of earths are of the negative kind. An earth is that fubftance, which is not foluble ; not fo heavy as metallic bodies, nor is capable of combuftion. Criteria fuch as thefe betray our very limited and imperfect knowledge. Cronftedt indeed mentions another mark, the malleability of earths ; but this obfervation may be applied to falts, phlogiftic fubftance, and the brittle metals. As to their form not being changed by a red heat, the fame can be faid of the vitriolated vegetable alkali, of metals that require a much greater degree of heat for their fufion, and of other foffils. Any expanfion of their bulk is fcarce perceptible to the eye, though a red heat is always fure to produce it, unlefs counteracted by the diffipation of fome volatile matter, as in clay, aerated lime, and other fubftances.

§ LIII.

* Vol. i. p. 26.
† Vol. iii. p. 228.

§ LIII. *Metals.*

PERFECT metals are easily distinguished by their opaque shining surfaces and specific weight. Their malleability, which Cronstedt considers as their peculiar character, is no general criterion; for we reckon almost as many brittle as ductile metals.

§ LIV. *Phlogistic Substances.*

A CERTAIN degree of levity, with as much phlogiston, loosely combined, as will occasion inflammation, is necessary to the constitution of all bodies denominated phlogistic. Solubility in oil is not a distinguishing property of this class; as that menstruum, though producing no effect on plumbago, yet acts violently on lead, copper, arsenic, and other metals.

§ LV. *Mixed Fossils.*

WHILE we are giving our attention to the distinct arrangement of the several classes, it will be easily seen that we mean to consider such fossils only as are in a state of purity; that is to say, free from every corruption by combination with the subjects of other classes, not necessary to their composition. Sulphurated metals, for example, belong to two classes; and we are

to

to determine from other data, to which they ought in preference to be adjudged. In like manner, aerated and fluorated lime, muriated filver, and fome others are to be confidered.

§ LVI. *Affinity of Foſſils.*

By the law of continuity, we may obferve a great affinity among the feveral claſſes of foſfils.

§ LVII. *Affinity of Salts with Earths* **and Metals.**

WE have already taken notice of the connexion of falts with earths, and we may add further to our remarks on this fubject, that burnt lime, by the intermedium of the matter of heat, acquires a folubility perfectly faline. The fame thing happens to ponderous earth, but not to magnefia. In all metals there lurks a certain acid peculiar to each, the nature of which we have as yet explored in arfenic only. Thefe metallic acids differ from all others in this refpect, that, when taken with proper proportions of phlogifton, they become metallic calces; but if faturated with that principle they are reduced to a perfect metallic ftate *, generating at the

fame

---

* Effays, vol. 3. p. 124.

fame time fulphur and aeriform fluids *. Moft phlogiftic bodies likewife, perhaps indeed all, contain an acid united in their very conftitution.

§ LVIII. *Affinity of Earths with Metals.*

EARTHS refemble the calces of metals in many of their properties; but in refpect to fpecific gravity, the faculty of colouring glafs, and their reduction to the metallic ftate, they are effentially different.

§ LIX. *Sulphureous Character of Metals.*

METALS in their perfect ftate are either metallic acids faturated with phlogifton, or a fpecies of metallic fulphur, which are fometimes very evidently fufceptible of inflammation, as zinc and arfenic. Gold and copper, when in fufion, afford fome appearance of flame, though faint, in a greenifh vapour; bright fparks are emitted from iron in a white heat; and tin alfo may be inflamed by a proper manner of operating.

§ LX. *Stones.*

IN the claffes already enumerated, all foffils are by no means included. Such as are compofed

---

* Effays, Vol. ii. p. 352.

fe l of heterogeneous fubftances, mechanically
mixed, and united in a vifible manner, and which,
for the moft part, conftitute the entire fummits
of mountains, are comprehended under one
name of *Petræ* or *Saxa.* Cronftedt has, with
great propriety, treated thefe feparately in an
appendix. The knowledge of thefe fubftances
is doubtlefs highly neceffary, and tends much
to the illuftration of phyfical geography ; but
they are not therefore to be confounded with
bodies more homogeneous, whofe combination
refting on chemical principles, is effected in the
way of folution.

§ LXI. *Organic Foffils.*

ORGANIC foffils are confidered by Cronftedt
in another appendix. Thefe fubftances are to
be treated as ftrangers from the animal or vegetable kingdom. They are diftinguifhed by an
organic ftructure, more or lefs imperfect; of
which, as long as they bear any marks, we are
to reckon them as foffils of a foreign fpecies.
The confideration of them is however in various
points of view, highly ufeful. They refemble a
feries of ancient coins in the teftimony they
bear to the convulfions and revolutions of our
globe, on which hiftorical monuments are wholly filent. From them we may learn the wide
extended fovereignty of **the** fea ; the changes

that

that fucceffive ages have wrought upon the furface of the earth ; and they difclofe to us what
animals inhabit the deep abyffes of the ocean,
and many other circumftances moft worthy the
attention and enquiry of philofophy.

§ LXII. *Volcanic Productions.*

THOSE burnt fubftances thrown out from the
mouths of volcanos, by a greater or lefs degree
of fubterraneous fire, Cronftedt has thought fit
to arrange in a third appendix. A general view
of them no doubt would be ufeful ; but there
are not wanting many reafons why, in my opinion, volcanic productions will not admit of a
feparate claffification. We know there are many who ftrenuoufly fupport the hypothefis, that
the whole foffil kingdom owes its origin to fire ;
for fuch as thefe, therefore, any diftinction will
be unneceffary. We have learned alfo, that
marks burned by fire into foffils are gradually
obliterated by the injuries of time ; becoming
firft obfcure, then equivocal, and at length being wholly deftroyed. Whatever limits, therefore may be drawn, they are in their very nature tranfient and perifhable. It is, and muft
be often exceeding difficult to determine whether foffils have derived their exiftence from folution, or from the effects of fire. According-
ly, to me it feems proper, to infert homogeneous
volcanic

volcanic productions into claffes agreeable to
their principles ; and that all thofe heterogeneous fubftances, whofe combination is vifibly mechanical, fhould be the fubject of the firft appendix.

Of Genera.

§ LXIII. *Arrangement of Genera.*

By the affiftance of claffes, all thofe foffils are
connected, whofe compofition, character, and
properties are perfectly fimilar. Genera require
a nearer affinity ; fpecies a refemblance ftill clofer ; and varieties muft correfpond in their internal habitudes only.

Foffils entirely homogeneous are of very rare
occurrence; as, for the moft part, two, three,
or more principles, enter into their compofition.

The more fimple their compofition, it follows,
they will be the eafier reduced to their natural
genera.

Let A and B be the proximate principles of
any foffil, let A be heavier than B, the compound A B, will be then referred to the genus
of A ; but this admits of various exceptions.

Suppofe B poffeffed of a generic difference,
and that it is no where found in a fingle ftate,
(for we do not here fpeak of artificial feparati-
on,)

on), but always united to A, or fome other matter, and ever inferior in weight in fuch combinations. According to the rule propofed above, the genus B fhould difappear entirely, and be altogether wanting in the genera of its own clafs, which is by no means confiftent with **a** natural fyftem.

Again, let us fuppofe B excels A in **the** intenfity of its properties, fo that B is only equal in weight to $\frac{A}{N}$, yet notwithftanding the qualities of B are clearly predominant in the compofition A B, that is, are much more confpicuous than thofe of the **lefs** ingredient A. Here again, unlefs I am deceived, we are to admit another exception.

If the cafes propofed under B and C obtain at the fame time, the exception receives a double confirmation.

Sometimes it feems neceffary to give a preference to the price of particular fubftances. Suppofe **A** B C an ore, whofe metal C, though of lefs weight than any other part of the mixture, yet in value furpaffes both B and A, fo that they are entirely neglected, and C only thought worthy the expence of metallurgic operations. In this cafe A B C is in fact the ore of C; but if the proportion of quantity were regarded, it fhould belong to the genus of A, and with great propriety, if a natural fyftem only is required.

We

We are not here to have any refpect to fictitious valuation. But as the arrangement of foffils is made with a view that our knowledge of them may be eventually ufeful and advantageous, it may feem to militate againft this defign, if we were to feek among the bafer kind for all thofe noble minerals, whofe intrinfic value can defray the labour and coft of eliquation.

The feveral cafes propofed ought not to be confidered as imaginary, as they each of them occafionally occur, and will be rendered more clear and intelligible by application in the following fections.

§ LXIV. *Genera of Salts.*

IN falts, we difcover two genera, by no means ambiguous; the acid, and the alkali. Chemiftry has not yet been able to extract their proximate principles; but, that they are different from, and oppofite to each other, there is not the leaft room to doubt.

§ LXV. *Acids.*

AN acid is eafily difcoverable by the tafte, by its property of changing to red the blue vegetable colours, and of effervefcing with aerated alkalis.

§ LXVI.

§ LXVI. *Alkalis.*

ALKALIS are diftinguifhed by a burning tafte, by their converfion of blue vegetable colours to a green, and by their powerful attraction for acids.

§ LXVII. *Salts not* **faturated.**

UNSATURATED combinations of acids and alkalis, enter the genus of the prevailing fubftance, unlefs any one fhould chufe to refer them rather to the imperfect neutral falts; which might be done not altogether without reafon, as the moft of them betray **an** excefs of either the one or the other ingredient.

§ LXVIII. *Whether neutral Salts are* **to be refer** *red to a diftinct Genus.*

IT may be queftioned whether an acid exactly faturated with an alkali fhould conftitute a diftinct and feparate genus? Or ought rather fuch a combination to be ranked under the acid, or the alkaline falts? If there is evidently an excefs of either of thefe principles, as in § 75, then, without doubt, it may be properly affigned to the genus of the exceeding principle; but, in all perfect neutral falts, the properties of acid and alkali are blended fo intimately by faturation, that all diftinction between them feems en-

Q   tirely

tirely to have difappeared. In this ftate of e-
quilibrium, then, it becomes a matter of indifference whether the preference be given to the
acid or the alkali. To the latter however I
fhould rather incline, as the moft convenient;
but I would not violently oppofe any one who
might think proper to refer them to the acid, or to
a diftinct genus. Quantity may in this cafe, in
fome meafure, affift our determination; but not
without irregularity : For, as the pure fixed alkali is faturated with a weight of acid lefs than its
own; fo, on the other hand, the volatile alkali
requires the acid to be heavier than itfelf.

§ LXIX. *Mixed Neutral Salts.*

It may happen, that the fame acid is partly
faturated with one alkali, partly with another;
and yet neverthelefs, thefe three are fo ftrongly
united by cryftallization, as to conftitute but
one peculiar falt. The falt of Seignette affords an
inftance of this fpecies of compofition; the
cream of tartar likewife faturated with volatile
alkali. That the fame alkali may be combined
with two acids, the union of cream of tartar
with the acid of borax fufficiently demonftrates.
In the foffil kingdom, indeed, we find none of
thefe triple falts; but they inform us what may
be done towards eftablifhing a general arrangement. The falt of Seignette, with the acid of
borax

borax, produces a quadruple falt ; and it is not unlikely, but that the induftry of future ages will difcover combinations of five principles, and per-haps of ftill more ; the difpofition and **order** of which may be determined by the character and quantity of the feveral ingredients.

§ LXX. *Analogous* **Salts.**

FOSSILS of the fecond and third clafs become true faline fubftances, by combination with any falt ; and in this condition they are banifhed from their original claffes. Salts, fuch as thefe, are called, analogous ; and according to the charac-ter of their bafes, are of two kinds, either earthly or metallic. Whatever imparts the faline na-**ture** ought to determine the genus.

§ LXXI. *Other Combinations of Salts.*

ALL earths almoft as well as **metals are not** only taken up by acids, but feveral foffils **befides,** of both claffes, are diffolved by alkaline falts, and fome even by neutral falts ; nay, it happens occafionally, that two double falts will unite in-to one, and form a falt of four principles. From fuch multiplied and various combinations pro-ceed, alkalis and acids charged with earths and metals ; double neutral falts, or falts of more principles, containing earths and metals ; double earthy falts united with double metallic falts,

Q 2   which

which, according as the faline matter is either the fame or different in each, generate triple or quadruple compounds.

§ LXXII. *Doubtful Genera of Salts.*

In the clafs of falts it often **happens, that** fome principles are never found **in a fingle and** independent ftate, but united always with others. Such are, for example, the nitrous, the muriatic, and arfenical acid. It may be doubted, therefore, whether thefe fubftances are to be confidered under their fimple genera. As, however, it does not feem improbable, that they were once free and uncombined, we are hardly authorized to exclude them ; though it may be, at the fame time, obferved, that they have never yet been found otherwife than in this ftate of combination. At all events, the inveftigation of fimple fubftances will throw light upon the feveral compofitions.

§ LXXIII. *Genera of Earths.*

Some genera of earths have hitherto refifted all attempts to reduce them into fimpler principles ; while others, by a proper analyfis, have difcovered two or more. The former are called *primitive*, the latter, *derivative* earths.

§ LXXIV.

§ LXXIV. *Primitive Earths.*

CRONSTEDT has eſtabliſhed nine primitive earths, but accurate experiments have ſince ſhewn that the greater number of them were compounded, ſo that the account is reduced to three only; the calcareous, ſiliceous, and argillaceous. We have however to add new earths, with which he was not acquainted, the terra ponderoſa and magneſia. We reckon therefore five primitive earths.

§ LXXV. *Of the common Origin of Earths.*

ALTHOUGH the powers of chemiſtry have not yet been able to decompoſe theſe five earths, the reduction of them all to one ſpecies, or, at leaſt, to a ſmaller number than the preſent, may poſſibly be the reward of future induſtry. I acknowledge myſelf of this opinion, and I think with ſome foundation. Clay, for example, is nothing elſe than calcareous earth, ſo ſtrictly combined with ſome unknown acid, that the ſeparation of them has hitherto been attempted in vain. No one certainly could have ſuſpected the calcareous baſe in the *lapis ponderoſus,* which has been demonſtrated by analyſis. In like manner, other ſubſtances may be inveſtigated. But until proper experiments ſhall have fully developed the nature of ſuch compoſitions,

Q 3

they

they muſt be, in reſpect to our knowledge of them, conſidered as primitive ſubſtances : For it **is** wholly inconſiſtent with the caution and diffidence of natural philoſophy to advance any poſition upon a bare poſſibility. Daily experience ſufficiently teaches, that thoſe things which at one time appear highly probable, may at another be diſcovered to be entirely unfounded.

§ **LXXVI.** *Reaſons why the Terra Ponderoſa* **ought to** *be referred to a diſtinct Genus.*

THE ponderous earth, on account of its great ſpecific gravity, is deſerving of particular attention, and leads us naturally to apprehend it to be of metallic origin. Other arguments alſo ſupport this hypotheſis. It is admitted, with the force of an axiom, that phlogiſticated alkali precipitates metallic ſolutions only : But if this alkali is dropped into a ſolution of acetated ponderous earth, it is immediately diſturbed, and a white powder is precipitated ; which, on examination, is found to conſiſt of that earth vitriolated, from the vitriolic acid inherent in the Pruſſian blue. If the powder is ſeparated by means of a filtre, and a new portion of acetated ponderous earth added to the liquid, on expoſing it to the fire, the ſolution, though clear before, depoſits another white powder, containing the ponderous earth united with the phlogiſtic alkali. The reſult is **the** ſame if the ponderous
earth

earth, faturated with the nitrous acid is treated in a fimilar manner: Therefore it feems rather to refemble a metallic calx than an earth, by thefe properties.

Among the metallic calces, that which arifes from lead correfponds with the ponderous earth in its weight, its white colour, and peculiar attraction for the vitriolic acid, by which that a-**cid is torn away from** alkaline falts ; **but there** is notwithftanding a remarkable difference between them. Acetated lead is difturbed wholly in the cold by phlogifticated alkali, and depofits a fediment, which neither is foluble in water, nor in the vitriolic acid ; but the acetated ponderous earth yields its genuine precipitate by heat only, and which is foluble both in the vitriolic acid and in boiling water. Befides, this earth has hitherto refifted all efforts to reduce it to a metallic ftate.

Therefore, although there may appear a confiderable affinity between the ponderous earth and a metallic calx ; yet, as long as it is incapable **of** reduction, its metallic nature is certainly not fufficiently demonftrated, and it muft ftill retain a place among the earths.

§ LXXVII. *Five Genera fhould be conftituted of the five primitive Earths.*

As we have enumerated already five primi-**tive earths,** they naturally become the heads of

Q 4   five

five diftinct genera. It is very rare, if ever, that
they are found in a fimple ftate, being either
combined with one or more of the other earths.
The moft eafy method, therefore, would be to determine the genus of every **fuch** compofition,
according to the heavieft principle ; but the
cafes before feparately ftated, in § lxiii, are often
objections to this plan.

§ LXXVIII. *Exceptions.*

WERE this rule once admitted, we fhould lofe
altogether the magnefian and argillaceous genera ; for, in the compofitious hitherto examined, into which thofe earths enter, the filiceous
has been always found to outweigh the others,
although, from their character and properties,
they had both the fuperiority. Common clay
contains above half its weight of filiceous earth,
fometimes above three fourths, and **yet** the argillaceous qualities are fo diftinct, that thefe compofitions are unanimoufly denominated argillaceous. The fame richnefs and pre-eminence of
quality, with refpect to the filiceous earth, are
found in magnefia, and other fubftances.

All earthy compofitions, therefore, may be
determined by the genus of **that** ingredient,
which exceeds the others in weight, unlefs it be
filiceous, and not equal to feven-eights of the
whole. In fuch cafes, the genus ought to be
afcertained

afcertained by whatever ingredient aproaches neareft in weight to the filiceous.

§ LXXIX. *Compounded Earths are not united me-chanically only.*

But perhaps, all earthy compofitions are nothing elfe than many fubtle mechanical mixtures? At the very firft view indeed there feems fome foundation for fuch an opinion; but a more minute invefligation furnifhes evidence of a clofer union conftructed on other principles. The earth of alum immerfed in lime-water, and entering into fo ftrict a combination with the lime as not to be feparable but by chemical art, teaches us, that among primitive earths mutual attraction has a real exiftence. Befides, **as** almoft all thefe mixtutes generally form cry-ftaline concretions, we have another proof, not only of the minutenefs of their particles, but of **an** union perfectly homogeneous.

§ LXXX. *Genera of Metals.*

In the third clafs we are to conftitute as many genera, as we have known diftinct metals.

§ LXXXI. *Encreafed within a few Years.*

At the beginning of the prefent century, eleven

eleven metals only were known; but it had fcarce grown forty years older, before the difcovery was made of platina, a noble and ductile metal, and of three or four others, that were not malleable, as cobalt, niccolum, magnefium, and fiderum, which laft has hitherto appeared to differ from all the reft *. The fifth in molybdena is not yet fufficiently explored, to determine whether it fhould be reckoned among thofe already known, or conftitute a new fpecies; and to the fixth, in the acid of the lapis ponderofus, we may apply the fame obfervation. Of thefe two, however, we are in hopes the character of the firft will be foon difplayed by the induftry of Mr Hielm. The genera of metals, therefore, of which we can be certain, amount to fixteen, or fifteen at leaft; and it is not unlikely that this number will be increafed by future difcoveries.

§ LXXXII. *Arrangement of mixed Metals.*

In fection lxiii. we have a queftion refpecting the genera of minerals containing two metals, the one of which is more valuable than the other, but in lefs quantity. Examples of fuch minerals we find in the golden pyrites, which
hold

---

* Meyer and Klaprothius have proved it to be iron joined to the phofphoric acid; and our author, convinced by their arguments, changed his opinion.

hold a fmall proportion of gold united with a
large proportion of iron ; among the galenæ,
that are far richer in lead than in filver; among
the copper pyrites, always producing more iron
than copper; and fo on of many others. According to fyftematic rules, the more valuable
and fcarcer metal, although it defray the expence of eliquation, fhould yet be referred to
the genus of **the** more abundant, though of lefs
eftimation. But if the ufe and aim of any fyftem
is confidered, there can be no doubt that the
preference fhould be affigned to the metal of
the higheft value. In fome degree, however,
the determination of this point may be **a mat-**
ter of indifference, provided no diftinct genus
is thereby deftroyed; a circumftance that would
probably affect the fiderite, in cafe it were de-
cided in favour of fuperiority in weight, as that
metal has never yet been found feparate from
iron ores, to which it always bears the fmalleft
proportion.

§ LXXXIII. *Genera of Phlogiftic Bodies.*

THE fourth clafs contains the feweft genera,
fulphur, petroleum, amber, and perhaps diamond.

§ LXXXIV. *Sulphur.*

SULPHUR is an inftance of the moft fimple compofition, confifting of two principles only, acid faturated with phlogifton.

§ LXXXV. *Petroleum.*

IN petroleum we difcover an union more complex ; a fmall portion of water combined, by means of an acid, with the principle of inflammability.

§ LXXXVI. *Amber.*

THE origin of amber is evidently from the vegetable kingdom, for, befides its peculiar acid and oil, we obtain the acetous acid by diftillation. The earthy refiduum may be confidered as a matrix.

§ LXXXVII. *Diamond.*

WITH regard to the diamond, I have hitherto found no place fo proper for it as this clafs. In a fufficient degree of fire, it is entirely confumed, and with an appearance of cloud or flame; and, in the focus of a burning lens it difcovers figns of a footy matter.

§ LXXXVIII,

§ LXXXVIII. *Pyrites and Molybdena do not con-*
stitute peculiar Genera.

I HAVE referred pyrites, or fulphurated iron **to**
the genus of iron. In like manner, molybdena,
which **is** nothing elfe than a metallic caix mineralized by fulphur, provided its genus were
known, ought to be afcribed to **the** clafs of metals. As to the foffil confidered **by** Cronftedt as
fixed phlogifton, and which he calls *brandertz*,
its compofition has not as yet been fufficiently
inveftigated.

§ LXXXIX. *Properly fpeaking, there is but one Ge-*
nus of phlogiftic Subftances.

In the ftrictnefs of language, all the genera
of this clafs might be reduced to one, as the
fame principle of inflammability prevails in each
of them.

§ xc. *Firft Appendix.*

In the firft appendix to the claffes, are treated
thofe foffils of various and mechanical combination, and which for the moft part is obvious to
the fight.

§ XCI.

§ xci. *Four Genera of Foffils mixed mechanically.*

ANSWERING this defcription, we have four genera only, which are denominated according to the clafs of the moft predominant ingredient in their compofition.

§ xcii. *Firft Genus.*

THE firft genus in which the faline character prevails occurs fometimes in the neighbourhood of volcanoes. In gypfum alfo other foffils intimately mixed are occafionally found. The fubftances likewife contained in natural waters may perhaps be referred to this genus. They are indeed held by water in folution, but their union is generally merely mechanical, of which the fixed principles are collected in the refidua, after the evaporation of the liquor.

§ xciii. **Second Genus.**

To the fecond genus we affign all thofe foffils in which the earthy principle abounds. Such are thofe placed by Cronftedt in his firft appendix under the name of *faxa.* Under this genus may be arranged feveral matrices of metals as well as of inflammable fubftances; for lithan

thrax,

thrax *, aluminous fchiftus, aluminous ore of La Tolfa, and many others, contain fome extraneous earthy matter, and in confiderable quantity.

§ xciv. *Third Genus.*

In the third genus, the metallic nature is predominant. It has been long obferved, that fome metals affect a difpofition to affociate with each other; fo that if one is difcovered, it may be properly conjectured that the other is not very far diftant. Relations fuch as thefe, as are obvious in this genus, are worthy attention and enquiry, as they promife no fmall advantage to the inhabitants of mountainous countries.

§ xcv. *Fourth Genus.*

In the fourth genus we meet with various mixtures of foflils, of which this ruling principle belongs to the laft clafs.

§ xcvi. *Diftinct and mixed Particles of Foffils.*

To this appendix likewife, the diftinct and mixed particles of foflils may conveniently be referred, inferting them under their proper genera, according to circumftances. Such, for example

---

* Pit coal.

example are the marles, moft of the common clays, mixed fands, and feveral others.

§ XCVII. *Four Genera* of organic Foffils.

LASTLY, Organic foffils are divided into four genera, as the diverfity of their nature fuggefts, whether they are found impregnated with and compofed of falts, earth, metals, or phlogifton.

§ XCVIII. *Fifth Genus of Cronftedt.*

CRONSTEDT adds a fifth genus, and perhaps with great propriety, in which are included all the dead remains of once living fubftances, which, by gradual putrefaction, have loft their original ftructure, though they ftill retain fuch ftrong marks of it as are not obliterated entirely but by the lapfe of many years. To this genus belongs the earth of deftroyed animals or vegetables.

§ XCIX. *Organic Bodies mineralized by Salts.*

THIS operation muft vary according to the nature of the fubftance. Bodies immerfed in a falt folution are fometimes penetrated by it, and indurated. In this manner the entire bodies of men, that had fallen by accident into the vitriolated water of the mine of Fahlun were found

after

after several years, so little changed to the eye, that the individual could be remembered by his countenance: In other respects however they were rigid like a statue, formed of saline matter. When exposed to the free air they began to crack. By a similar process, no doubt, even softer substances may be so hardened, as to preserve their structure a long time, exempt from putrefaction.

§ c. *Bodies impregnated with Bitumen.*

IN like manner organic bodies, impregnated with bituminous matter are enabled to preserve themselves from decay, and retain their figure and structure.

§ **CI.** *Petrifaction of organic Bodies.*

NEITHER the bodies of animals nor of vegetables can be wholly penetrated by stony particles. The harder parts only, as the bones, shells, external covering, roots, woods, fruit, and similar substances, are liable to this change; which, if I mistake not, proceeds in the following manner : At first, the parts of softest texture putrefy, and leaving several empty spaces, through which water loaded with earthy particles passes, and in its course depositing them, the vacuities are at length filled by their gradual accumulation. Then follows the destruction of the more firm consistence, to be penetrated in the same order.

R   If

If the later depofitions differ in their colour and
properties from thofe of an earlier date, yet the
original organic ftructure is beautifully difplayed
by fmooth and polifhed fections of the different
bodies. All the particles, however, of the bodies fo deftroyed are not always carried off; for
it often happens in diftillation, that fuch are ex-
pelled as fhew figns of an organic conftruction.

§ cii. *Organic Bodies penetrated with metallic
Particles.*

The moft fubtle metallic molecules, that can
poffibly be carried along by water, may in the
fame manner penetrate and change the harder
organic parts.

§ ciii. *Nuclei.*

From the fubftances already defcribed, nuclei have, with great propriety, been confidered as quite diftinct. They are produced by two
different proceffes. Any body poffeffing a fhell
or firmer covering, and depofited in a foft ftratum, is gradually attacked in its flefhy parts
and foft inteftines, which are either wholly deftroyed, or contracted by exficcation; fo that
room being made in this manner for the particles flowing in, the fhell is at length filled with
a nucleus, bearing the marks of its internal furface. If a body is involved in fediment, and
after the exficcation of the ftratum is any way
deftroyed

deftroyed or carried off, a nucleus will be formed in the cavity, defcribing its external fea-
tures.

§ CIV. *Remaining Impreffions of organic Bodies.*

IN any foft fubftance, impreffions are left by
cockles, fnails, infects, fifhes, and other fmall a-
nimals of the firmer kind, either of their external furface, their bones, or fkeletons.

CV. *Ofteocolla.*

IN particular foils, living roots are by degrees
covered with fo hard a cruft, as to prevent the
abforption of the neceffary juices. When a vegetable attracts moifture every where in the
neighbourhood of its root, the fubtile, calcareous, argillaceous, filiceous, and even ochreous
molecules, that accompany it, produce this effect. The fluid in which they were borne being
abforbed by the roots, they fix themfelves on
the furface, and there forming a covering imper-
vious to water, the roots decay, putrefy, and
leave this cruft, which is commonly called *ofteo-*
colla.

§ CVI. *Incruftated organic Bodies.*

WATERS loaded with earthy particles frequently cover with a cruft, reeds, fmall bran-

ches,

ches, and other fubftances immerfed in them, without any alteration of their original form.

OF THE DIFFERENT SPECIES.

§ CVII. *Specific Characters of Salts.*

SPECIFIC characters are to be determined by the difference in the nature of thofe fimple falts, which art has not been able to compofe from their principles. Of thefe, two **diftinct** genera only are known; the acid and the alkali already mentioned.

§ CVIII. *Species of Acids.*

THE genus of acids is very extenfive. **The** vitriolic, nitrous, and muriatic, have been extracted from foffils for many ages paft; but the difcovery of others differing evidently from thefe has been made within a much later period. The acid **of fluor,** borax, arfenic, fiderite, molybdena, and lapis ponderofus, are of this defcription *.

§ CIX. *Vegetable Acids.*

WE have the profpect as yet of a more extenfive field in the acids of the vegetable kingdom.
Befides,

* For metallic acids, fce Effays, v. iii.

Befides the acetous, which was **the** only one formerly known, it has produced to us already the acids of fugar, forrel, tartar, benzoin, **citron,** amber, and feveral others.

§ cx. *Animal Acids.*

The animal kingdom is the pooreft of the three; for except the acid of ants, and of fat, we know of none other proper to it, although, without doubt, it contains many highly deferving of notice. As for example, the acid which the larva phalænæ vinulæ of Linnaeus throws **out in its** defence, **clear as** water, and colourlefs, which refembles **the** concentrated acetous acid in fmell and tafte, coagulates blood, and thickens fpirit of wine; reddens blue paper for a fhort time; but the original colour returning afterwards, affords proof of its great volatility [*]. The fcarcity of this very fingular liquor has perhaps delayed fo long its further inveftigation.

§ cxi. *Acids common to feveral Kingdoms of Nature.*

Other acids are common to all the kingdoms of nature, as the *phofphoric*, which had been falfely affigned to the animal kingdom alone; but which has been found, though rarely, in the foffil [†], **and** in **great** plenty in the vegetable kingdom.

R 3

[*] Oeuvres de M. Bonnet, v. iii. 8vo. p. 28.
[†] Effays, vol. ii. page 426.

kingdom. Under this head we may arrange the aerial acid.

§ CXII. *Great Number of Acids.*

If we confider, that probably the exiftence of all metals depend upon their peculiar radical acids; that vegetables evidently contain a number of unknown acids; and that, perhaps, the fame may be faid of animals alfo; we have reafon to wonder at the abundance and variety of this fubftance, and to fet a high value on its utility and importance in the œconomy of nature.

§ CXIII. *Species of Alkaline Salts.*

The extent of the other genus is confined within very narrow limits. For a long time three fpecies only of alkaline falts were known; two of which could bear a flight ignition, and were therefore denominated fixed; while the other was diftinguifhed by its volatility.

§ CXIV. *Fixed Alkalies.*

Of the fixed alkalies the one feems to prevail in the vegetable, and the other in the mineral kingdom; from which they both derive their names.

§ CXV.

§ CXV. *Neutral Salts.*

SALTS formed by the exact faturation of acids with alkalies amount to fixty double fpecies, on the fuppofition that the acids do not exceed twenty in number. A confiderable part, however, of the combinations of thefe are as yet unknown, or at leaft but imperfectly examined.

§ CXVI. *Imperfect double Salts.*

MANY imperfect double falts have been difcovered. The acids of vitriol, arfenic, tartar, and forrel unite in excefs with the vegetable alkali; and the acids of vitriol and tartar with the mineral alkali. The labours of pofterity will probably add a greater number. Borax retains an excefs of alkali; and the arfenicated mineral alkali likewife is capable of a fimilar combination.

§ CXVII. *Triple Salts.*

THE falt of Seignette, and tartar faturated with volatile alkali, furnifh examples of the neutral triple falts.

§ CXVIII. *Imperfect Triple Salts.*

AMONG the triple imperfect falts, we know of the union of tartar with the acid of borax. Here is an excefs of acid.

§ CXIX. *Quadruple Salts.*

TARTAR and borax combined, are an in, ftance of the quadruple falts.

§ CXX. *Species of analogical Salts.*

EARTHS and metals, although fingly they refufe every combination with water, yet by the admixture of a falt they become for the moft part foluble, and are then called analogical falts.

§ CXXI. *Species of double perfect earthy Salts.*

FOUR primitive earths uniting with twenty acids, produce eighty double perfect earthy falts; that is falts compleatly faturated. The fifth earth, the filiceous, is foluble in the fluor acid only.

§ CXXII. *Double imperfect earthy Salts.*

OF all the double imperfect earthy salts, with an excess of acid, the salt of alum is the most conspicuous.

§ CXXIII. *Triple earthy Salts.*

THE principle triple compounds, are the volatile alkali, either vitriolated or muriated, and magnesia, with which even nitrated lime readily unites.—Vitriolated magnesia combines with clay; and both the vegetable and mineral alkali saturated with the acid of fluor, admit an union with siliceous earth·

§ CXXIV. *Earthy alkaline Salts.*

FIXED caustic alkalis, I know for certain affect no other earths than the argillaceous and siliceous. No triple alkaline salts have as yet been discovered.

§ CXXV. *Species of metallic salts.*

ANALOGICAL metallic salts are by far the most numerous. From a combination of the sixteen metals with the twenty acids, we obtain three hundred and twenty double salts; but

but which can be scarcely so perfectly saturated, as that there should not be some small excess of acid.

§ cxxvi. *Metallic Salts, with an Excess of the metallic Base.*

THERE are some instances also of the union of metals and acids, highly deserving of notice, in which the excess is on the part of the metal. To this head we refer the turpith mineral, and red precipitate of Mercury, which though ever so well washed, yield a small quantity of a-cid on distillation. The same remark applies equally well to the pulvis algarothi. Mercurius dulcis retains its metal partly calcined and part-ly perfect *; and nitrated silver, in like manner can take up a portion of silver, without dephlo-gisticating it. Muriated copper, deficient in its acid, constitutes a peculiar salt hitherto undis-cribed.

§ cxxvii. *Triple metallic Salts.*

WE have long been acquainted with a con-siderable number of metallic triple salts, that are not separable but by decomposition. Of this description are the combinations of tartar with iron and antimony ; of the vitriolated ve-getable

* Scheele in Actis Stockh.

getable alkali with iron; of the muriated vegetable alkali with platinum ; of the vitriolated volatile alkali with copper; of the muriated volatile alkali with platinum, quickfilver, copper, and iron; of vitriolated and acetated quickfilver with iron ; of vitriolated iron with magnefium, with copper, and with zinc.

§ cxxviii. *Quadruple metallic Salts.*

THE quadruple metallic falts are formed by the union of fal ammoniac with nitrated iron, with nitrated copper, and with boracic quickfilver; of the vitriol of iron, likewife, with the vitriols of copper and zinc together.

§ cxxix. *Alkaline metallic Salts.*

MOST of the alkalis alfo combine readily with metals, efpecially the volatile alkali; which fometimes forms beautiful cryftals, with a metallic bafe, as with filver and copper. The numerous family of thefe falts are deferving of much greater attention than has ever yet been paid to them.

§ cxxx. *Synopfis of Salts.*

FROM what has been faid, I am of opinion there can be no doubt of the extenfive influence

and

and variety of the clafs of falts, in which we have here confidered all thofe prepared by art, as well as thofe produced by nature. In favour of the halurgic fyftem, I fhall fubjoin a table, prefenting at one view all the chief varieties, with which I am acquainted. A greater number of proper experiments would certainly add many more to the account.

S A L T S.

Properly fo called	Simple	{ Acid { Alkali.
	Double	{ Neutral { Imperfect.
	Triple	{ Neutral { Imperfect.
	Quadruple	{ Neutral { Imperfect.

Analogic	earthy	with an acid	{ double { triple	{ Perfect { Imperfect
		with an alkali	double	
	metallic	with an acid	{ double { triple { quadruple	{ Imperf. with excefs of acid { Imperf. by defect of acid.
		with an alkali	double	

§ cxxxi.

§ cxxxi. *Species of Earths of a double Character.*

In the clafs of earths different fpecies frequently occur, poffeffing two characters. To the firft belong the faline earths; which, on account of the limits before affigned to them, are not reckoned in the clafs of falts, although they refemble them in their nature, and conftitute but an imperfect fpecies of earths. Of thefe fubftances, however, a few only are known, § 51.

§ cxxxii. *Mixed Species of Earths.*

GENUINE fpecies of mixed earths **are produced** by the intimate union of two or more. **Of** the exiftence of fuch an union we have clear **evidence**, in § 90.

§ **cxxxiii. On what Arguments their Diverfity is founded.**

Not the quality and number **only of the ingredients**, but even their **relative weights imply a** fpecific **diverfity.**

§ cxxxiv. *The Neceffity of confidering the Proportion of every Part.*

IN the Sciagraphia Regni Mineralis, lately publifhed

publifhed, I have overlooked the mutual proportions; but, on further reflection, I find the confideration of them abfolutely neceffary.

§ cxxxv. *Method of inveftigating the feveral Species of Earths.*

In order to determine with accuracy the fpecies of earths, which hitherto feem to have refted on no very certain foundation, it will be requifite to explain carefully this doctrine. Let the five primitive earths be indicated by five initial letters, the ponderous by *p*, calcareous by *c*, magnefian by *m*, argillaceous by *a*, and filiceous by *s*.

§ cxxxvi. **Continuation.**

At firft we will attend to the character only and number of principles; and, by means of the doctrine of combinations, it will be eafy to afcertain how many fpecific confociations can arife from thefe five letters.

For example, *p, c, m, a,* and *s,* can produce no **more** than ten double fpecies—

pc,   pm,   pa,   ps,
cm,   ca,   **cs,**
ma,   ms,
as.

Of

Of triple species we have as follows :—

*pcm, pca, pcs, pma, pms, pas,
cma, cms, cas,
mas.*

Quadruple :—

pcma, pcms, pcas, pmas.

Lastly, One quintuple only :—

pcmas.

In this manner, from the whole class of earths, besides the five simple species, containing the primitives alone, we can obtain but twenty-six different combinations; which, together with the five simple, amount in all to thirty-one.

§ CXXXVII. *Why this Method is imperfect.*

In this plan, however, the number of the species is too much limited, and our conclusions liable to error. It will easily appear that *pa,* for example, must be separated ; for the character of the mass, with an excess of ponderous earth, will be by no means the same as with an excess of clay. In like manner *pac* should be referred to three distinct genera, according as the first, the second, or the third principle bear the greatest share in the composition, (§ 78.). The same, indeed

indeed, will be obferved in whatever formula is employed. Therefore it is neceffary, together with the number of the principles, to confider the weight of each.

§ CXXXVIII. *In what Manner can this De-fect be fupplied or corrected.*

THAT they may be all fymbolically defigned, **and** rendered obvious to the fenfes, a certain lo-**cal value muft** be affigned to every letter; fo **that whatever** principle occurs firft in combina-tion, that fhould be underftood to be the heavi-eft of the whole mafs: Every intermediate prin-ciple will yield to the preceding one, but ex-ceed thofe that follow it, and the laft of all will be of the leaft importance.

§ CXXXIX. *Enumeration of double Species.*

ACCORDING to this fyftem then we fhall have twenty double fpecies :

pc,   pm,   pa,   ps.
cp,   cm,   ca,   cs.
mp,   mc,   ma,   ms.
ap,   ac,   am,   as.
ſp,   ſc,   ſm,   ſa.

§ CXL.

§ CXL.   *Enumeration of triple Species.*

EACH of the five letters in forming triple compofitions, may be arranged in twelve different ways.   Five multiplied by twelve, therefore produce fixty fpecies as follows :

pcm, pca, pcf, pma, pmf, pmc, paf, pac, pam,
  pfc, pfm, pfa.
cpm, cpa, cpf, cmp, cma, cmf, cap, cam, caf,
  cfp, cfm, cfa.
mpc, mpa, mpf, mcp, mca, mcf, map, mac, maf,
  mfp, mfc, mfa.
apc, apm, apf, acp, acm, acf, amp, amc, amf,
  afp, afc, afm.
fpc, fpm, fpa, fcp, fcm, fca, fmp, fmc, fma,
  fap, fac, fam.

§ CXLI. *Quadruple Species.*

As the double fpecies amount to twenty ; and thefe, with the remaining three letters can be combined in fix different ways, in the quadruple fpecies, it will be eafily feen, that fix times twenty, or one hundred and twenty, will exprefs the amount of this divifion.

S

*pcma, pcam, pcmf, pcfm, pcfa, pcaf, pmac,
pmca, pmaf, pmfa, pmcf, pmfc, pacm, pamc,
pacf, pafc, pamf, pafm, pfcm, pfmc, pfac,
pfca, pfam, pfma.*

*cpma, cpam, cpmf, cpfm, cpaf, cpfa, cmpa,
cmap, cmpf, cmfp, cmaf, cmfa, capm, **camp**,
capf, cafp, camf, cafm, cfpm, cfmp, **cfpa**,
cfap, cfma, cfam.*

*mpca, mpac, mpaf, mpfa, mpcf, mpfc mcpa,
mcap, mcpf, mcfp, mcaf, mcfa, macp, mapc,
macf, mafc, mapf, mafp, mfcp, mfpc, mfap,
mfpa, mfac, mfca.*

*apcm, apmc, apmf, apfm, apcf, apfc, acpm,
acmp, acmf, acfm, acpf, acfp, ampc, amcp,
ampf, amfp, amcf, amfc, afpc, afcp, afpm,
afmp, afcm, afmc.*

*fpcm, fpmc, fpam, fpma, fpca, fpac, fcpm,
fcmp, fcam, fcma, fcpa, fcap, fmca, fmac,
fmpa, fmap, fmcp, fmpc, fapc, facp, facm,
famc, famp, fapm.*

§ CXLII. *Quintuple Species.*

THE triple species being fixty in **number,**
(§140.) and each of thefe admitting **of two**
changes only with the other two letters, it fol-
lows

lows, that, under this head, we may reckon one
hundred and twenty species.

pcmaſ, pcmſa, pcamſ, pcaſm, pcſam, pcſma,
pmcſa, pmcaſ, pmaſc, pmacſ, pmſca, pmſac,
pamſc, pamcſ, paſmc, paſcm, pacſm, pacmſ,
pſcma, pſcam, pſmca, pſmac, pſamc, pſacm.

cpmaſ, cpmſa, cpaſm, *cpamſ, cpſam, cpſma,*
cmpaſ, cmpſa, *cmapſ, cmaſp, cmſpa, cmſap,*
camſp, campſ, *capmſ, capſm, caſpm, caſmp,*
cſmpa, cſmap, *cſpma, cſpam, cſapm, cſamp.*

mpcſa, mpcaſ, mpacſ, mpaſc, mpſca, mpſac,
mcpaſ, mcpſa, mcapſ, mcaſp, mcſpa, mcſap,
mapcſ, mapſc, macpſ, macſp, maſcp, maſpc,
mſpca, mſpac, mſcap, mſcpa, mſacp, mſapc.

apcmſ, apcſm, apmcſ, apmſc, apſcm, apſmc,
acpmſ, acpſm, acmpſ, acmſp, acſpm, acſmp,
ampcſ, ampſc, amcpſ, amcſp, amſpc, amſcp,
aſpcm, aſpmc, aſcpm, aſcmp, aſmpc, aſmcp.

ſpcma, ſpcam, ſpmca, ſpmac, ſpacm, ſpamc,
ſcpma, ſcpam, ſcmpa, ſcmap, ſcamp, ſcapm,
ſmpca, ſmpac, ſmcpa, ſmcap, ſmacp, ſmapc,
ſapcm, ſapmc, ſacpm, ſacmp, ſampc, ſamcp.

ĝ CXLIII.

§ CXLIII. *Amount of the Species.*

IF the primitive earths are five in number, then the preceding paragraphs. exhibit the formulæ of all thofe fpecies that can poffibly a-rife from their various combination; and to which, adding the five fimple earths, we fhall find the amount to be thus, $5+20+60+120+120=325$, the amount of the whole.

§ CXLIV. *Further Explanation of the Formulæ.*

I HAVE fo contrived thefe formulæ as to make it evident to what genus every combination is to be referred. — The firft letter determines the character of that genus, *s* only excepted; as, though it exceeds in weight, yet its other qualities do not always prevail, (§ 89.)

If at any time the number of the primitive earths is diminifhed, whether by decompofing them into others more fimple, or by difcovering them to be of a metallic nature, yet the fame formulæ may be preferved after making the neceffary correction.

For example, Suppofe *p* were referred to the third clafs, the quintuple formulæ, (§ 142.) would then become quadruple, that feries being deftroyed entirely where *p* begins, and from all the others would it be taken away. In this cafe, we we lofe the whole of the firft genus, and the fame formulæ are repeated four times in each of the remaining genera, and conftitute one fpecies only; fo that $\frac{12}{2}=6$ fpecies is of each genus and $4\times6=24$ the number of all the quadruple fpecies.

Let us take another example, and remove altogether a, the formulæ of that genus are immediately annihilated, and the eighteen in the three other genera are reduced to $2\times3=6$.

In the fame manner, that the corrections are made in the formulæ of the laft order, can they be applied to thofe preceding. For it is evident that in reducing quadruple to triple fpecies, it is impoffible when p is deftroyed, that the remaining feries fhould be quadruple, and are therefore to be removed entirely.

Let n reprefent the number of primitive earths, and the number of the double fpecies be expreffed by $n. n.—1.$ of triple fpecies by $n. n—1. n—2$, of quadruple fpecies by $n. n—1. n—2. n—3$, and that of the laft order by $n. n—1. n—2.—n—n—2.$

§ CXLV. *Species of Metals.*

HAVING determined thefe points, we now proceed to the third clafs, in which, on account of the greater number of genera, we fhall find the fpecies alfo to be far more numerous.

Metals

Metals occur generally either complete, mineralized, or deprived of their phlogiſton.

§ CXLVI. *Native Metals.*

WHATEVER poſſeſſes a complete metallic form, is denominated native.

Into this ſtate no heterogeneous ſubſtances are admitted, unleſs they are perfectly metallic. Hence ariſe various ſpecies;—the metal native and ſimple;—combined with ſome other;—or with ſeveral together. Native ſimple metals are very **rare,** and, as far as I know, have never yet been diſcovered perfectly pure.

Moſt metals are occaſionally found native, as gold, platinum, ſilver, quick-ſilver, copper, biſmuth, niccolum, arſenic, cobalt, and antimony; but ſcarce any one of them occurs quite pure. Gold is mixed with ſilver or copper; ſilver with gold or copper; platinum with iron; niccolum and cobalt with arſenic as well as iron; antimony with iron or zinc; and further experiments will without doubt diſcover other combinations.

The exiſtence of native lead, **iron tin, and** zinc has been always much queſtioned **by many.**

Magneſium and ſiderite have never yet been found in a native ſtate.

§ CVLVII.

§ CXLVII. *Mineralised Metals.*

A MINERALISED metal appears to me to be a metal intimately united with fome foreign fubftance that deftroys more or lefs the genuine metallic form.

§ CXLVIII. *Mineralifing Subftances*

SUCH are fulpher and acids.

CXLIX. *Metals mineralifed by Sulphur.*

SULPHUR can be directly united with all the metals, except gold, platinum, and zinc; and thefe mineralifations are found in the bowels of **the earth.** Sulphurated tin alfo occurs in Siberia *.

Some mineralizations are affected, both as to character and appearance, according to the quantity of fulphur. Tin, combined with twenty hundred parts of fulphur, forms a mineralifation, white and fibrous; but, with twice that proportion, the compound is micaceous, and of the colour of gold.

Sulphur acting on perfect metals feparates a portion of their phlogifton; and is even capable of uniting with many calces likewife.

S 4 The

* Effays, vol. iii. p. 158.

The combination of gold with fulphur, by the intermedium of iron; is not yet made fufficiently evident; for that which is found in pyrites feems to be rather mixed than diffolved; as in a folution of pyrites, in the nitrous acid, the gold is depofited in molecules, not in powder, but differing from each other both in fize and figure *.

As to zinc, that metal appears in the pfeudo galena to be joined † with fulphur by means of iron.

§ CL. *Mineralifing Acids.*

OF mineralifing acids there are feveral, as the vitriolic, muriatic, phofphoric, aerial, and probably the arfenical.

§ CLI. *Vitriols.*

VITRIOLS of copper, iron, and zinc, are the fpontaneous productions of nature. Combinations of the fame acid with lead, niccolum, and cobalt, are likewife fometimes found ; and they feem generally to be the refult of decompofed mineralifations.

§ CLII.

* Effays, vol. ii. p. 412.
† Ibid. p. 329, and 336.

§ CLII. *Metals mineralifed by the Muriatic Acid.*

THE muriatic acid is more rarely found united with metals. As yet it has not been difcovered in any other than filver, quickfilver, and
copper. The two firft contain with it the vitriolic acid **likewife** *.

§ CLIII. *Metals mineralifed by the Aerial Acid.*

THE aerial acid is often prefent in calciform
metals. We meet with it in lead, copper, iron,
and zinc. Of its connexion with other metals
we have no certain intelligence.

§ CLIV. *Metals mineralifed by the Phofphoric Acid.*

OF all the acids, that of phofphorus is the
fcarceft, and has hitherto been found with a
fpataceous kind of lead only.

§ CLV. **Metals mineralifed by the Arfenical Acid.**

THE arfenical acid, if I miftake not, is the
true menftruum of the red cobalt, that is fometimes beautifully cryftallifed. It is certain, that
a red colour is owing to an acid, and that, from
all the experiments as yet made, no other has
been difcovered.

§ CLVI.

* **Woulfe, Philof. Tranf.**

§ CLVI. *The different Species of Metals admit of almost numberless Variations.*

WHOEVER confiders, that we are acquainted already with fixteen metals, and that of thefe the greater number of the perfect can be in feveral ways combined together, as well as thofe mineralifed by fulphur and various acids, will naturally expect that, by means of accurate analyfes, many more fpecies might be difcovered, which have as yet probably efcaped the refearches of the laborious philofopher. Were we to purfue the plan applied to the earths, (§cxliii.) the number would be really aftonifhing; but I am almoft of Pliny's opinion, who fomewhere confeffes: " Mihi contuenti fefe perfuafit re-" rum natura nihil incredibile exiftimare de ea." Formulæ, indeed, point out to us what may be done; but whether, and where, they are employed, muft be learned from a faithful analyfis; which affifts us, befides, better to underftand thofe of them that prefcribe the true limits to our inveftigations.

§ CLVII. *Spiecies of Phlogiftic Subftances.*

THE fourth clafs is exceedingly poor both in genera and fpecies.

§ CLVIII.

§ CLVIII. *Species of the Diamond.*

WE are acquainted with many differences of the diamond, but with none that are specific.

§ CLIX. *Species of Sulphur.*

THE species of sulphur are distinguished by the diversity of their acids, and we know of two only; the common formed by the vitriolic acid, and plumbago, containing the aerial acid saturated with phlogiston.

§ CLX. *Species of Petroleum.*

THE varieties of petroleum, in colour and tenuity, depend for the most part on the degree of exsiccation, and on the matrix or heterogeneous substances mechanically mixed with it; so that they can be considered but seldom as specific. Exsiccation produces a mass thick and tough, or solid and dry.

§ CLXI. *Amber.*

THE same observations nearly will apply to amber. In respect of transparency and colour, we meet with many varieties in the European species.

The

The Indian fpecies agrees in all thing with the European, except its being fofter, and wanting the volatile falt*, which laft circumftance feems to eftablifh a fpecific difference. Copal, commonly fo called, is to be diftinguifhed from the gum refin of that name fold by the apothecaries.

§ CLXII. *Origin of Phlogiftic Subftances.*

DIFFERENT opinions are maintained by philofophers, refpecting the origin of phlogiftic fubftances. Some contend, that thefe bodies are proper to the foffil kingdom; while others, probably with more reafon, afcribe them to thofe organic fubftances which abound in various oily and fat juices, and are not fo much affected by time, as they are gradually changed in the bowels of the earth by neighbouring pyrites and other foffils, until they acquire a bituminous quality. Heterogeneous fubftances enclofed within them are evident proofs of original fluidity. The different degrees of purity of naptha, coagulation performed by time, acids, or other media, and various circumftances befides in the great laboratory of nature, all influence the denfity, colour, clearnefs and other properties.

As to ambergrife, Aublet infifts, that it is the juice of a tree growing in Guiana, and there called

* Lehman, Chem. Schrift.

called Cuina. He says, that after heavy rains, large masses of it are washed into the rivers. The specimens examined by Roffelle are said to resemble ambergrise in their odor and principle qualities *. Long ago, Rumphus makes mention of a tree called nanarius, containing a juice similar to ambergrise. Lately, however, in England an opinion has obtained, that this substance is the excrement of a cetaceous fish. Observations made on the physeter macrocephalus, (the spermaceti whale) have given rise to this Idea, as the excrement in the intestines of that animal, is found on dissection perfectly hardened, and containing the beak of **the repia octopodia, on which it feeds, and in every** respect resembling the ambergrise of commerce.

§ CLXIII. *Species of Fossils mixed mechanically.*

OF fossils mechanically mixed, that fall under consideration in the first appendix, we have constituted four genera only, (§ 91.) their species, however, are numerous.

§ CLXIV. *The several Species expressed by the Formulæ of Letters.*

LET *s* denote salt, *t* earth, *m* metals, and *i* phlogistic substances; and let the same local value

---

* Hist. des plantes de la Guyane, 1774.

value be affigned to thefe letters as in the foregoing examples, (§ 138.) and we fhall obtain the
following double fpecies.

st, sm, si.
ts, tm, ti,
ms, mt, mi.
is, it, im.

Triple fpecies.

stm, sti, smt, sit, smi, sim,
tsm, tsi, tms, tis, tmi, tim,
mst, msi, mts, mis, mti, mit,
ism, ist, ims, its, itm, imt.

Quadruple fpecies.

stmi, stim, smti, smit, sitm, simt,
tsmi, tsim, tmsi, tmis, tims, tism,
msti, msit, mtis, mtsi, mits, mist,
istm, ismt, itsm, itms, imst, imts.

§ CLXV. *Continuation.*

WE are, however, **not** rafhly to conclude **that**
all the fpecies are exhaufted in thefe formulæ;
for every letter may be varied in many ways,
according to the diverfity of the feveral fpecies.
For example, *t* can be multiplied more than
325 times, (§ **131,** 143). *I,* indeed, prefents
but

but few variations, and *s* likewife; as the number of the falts proper for thefe mixtures, is exceedingly limited; but *t* furpaffes even *m*,
(§ 156); fo that we have here another occafion
of admiring the exhauftible ftories of nature.

§ CLXVI. *The Pofition and fituation of mixed
Foffils.*

It is by no means to be expected; that every
fpecies of thefe mixed foffils, which to me appear
to be *petræ*, fhould be equal to the production
of huge mountains. The greateft number of
them have hitherto been found in veins or fmall
ftrata only; many of which, though of different
characters, when combined, give birth to rocks.
The fame may be faid of the feparate particles,
which, in the aggregate, form large and continued ridges of hills. But thefe almoft always
fpring from the ruins **and** decompofitions of
mountains.

§ CLXVII. *Species of organic Foffils.*

Organic foffils conftitute four genera, (§ 97.);
but the feveral fpecies of foffils, whether poffeffing an organic form only, or with it an organic
ftructure, are diftinguifhed by fpecific marks.

§ CLXVIII.

§ CLXVIII. *Species of organic Foffils mineralifed by Salts.*

ORGANIC foffils, penetrated with faline matter, are but feldom found. Gypfum, indeed,
fometimes contains the lefs perifhable remains
of animals and vegetables ; but thefe fubftances
are fcarce ever found quite gypfeous. Entire
animals are occafionaly to be met, filled with vitriol, (§ 99.) and ftill oftener the harder parts of
vegetables, or their roots, feem to refift putrifac-
tion by the means of this falt.

§ CLXIX. *Earths.*

THE fecond genus, comprehending earthy
foflils, is by far the richeft. Innumerable calcareous nuclei of fhell fifh and marine infects
daily occur in calcareous ftrata. Sometimes, an
a imal covering, or fhell, which was before calcareous, being changed in its internal texture
only, become fpataceous.

Argillaceous nuclei of marine animals are common in aluminous fchiftus, but very rare in any
other bed. Frequently the covering of the a-
nimalcule ftill remains.

Marine exuviæ are obvious in marle alfo. If
lime predominates, often the fkeletons alone of
the fifh are feen. Of Ofteocolla we have already fpoken fufliciently, § 105.

Siliceous

Siliceous nuclei frequently fill entirely the internal cavity of organic foffils, and fometimes
even the fame matter furrounds their external
furface. I am in poffeffion of an echnites, the
fhell of which is filled with common flint, and
fhews upon the furface of the nucleus all its natural inequalities; the fhell itfelf, however is
calcareous and fpataceous, although it was im-
beded in filiceous earth on both fides. Small
fhells occur fometimes in jafper, but very rarely *,
and are not more frequent in petrofilex.

Organic bodies, themfelves alfo are found
penetrated with filiceous matter. Siliceous petrefactions of the trunks of trees are often dif-
tinctly marked with the growth of every year.
Siliceous mufcles and cockles alfo frequently
occur, and fmail corals even are fometimes clearly to be diftinguifhed in common flints.

I have feen the marks of leaves accurately
expreffed in quartz, and the epitomium of Blan-
kenburg is often quartofe.

Nuclei of fand are fometimes to be met with;
but the figure of their furface is generally fo obfcure, that it is very difficult to determine from
what organic body they were produced.

In the fand pit at Maeftricht was found not
long ago the fkeleton of a crocodile, fome teeth
of which were fent to me.

T

§ CLXX.

* Ferber in Epift. de Italia.

§ CLXX. *Species of organic Foffils impregnated with metallic Particles.*

VERY few metals affume an organic form. The calx of iron, but flightly cohering, or concreted like a ftone, penetrates roots, wood, and even whole trees, preferving ftill the fibrous **texture,** which may fometimes be fcraped with **the nail.**

Pyrotaceous iron, indeed, now and then forms nuclei ; but it commonly adorns the organic ftructure with lines or little fpots, and feldom occupies it entirely.

Copper, in the form of a calx is fuppofed frequently to enter into bones and teeth, giving them a blue colour, efpecially after they are calcined. This colour, however, is often owing to iron.

Pyritaceous copper alfo refembles the anomia in the magnet of Iarlfberg in Norway, and fifhes in feveral places.

Spots of native gold or filver **are fometimes feen** on the furface of foffil fhells.

The grey ore of filver at Frankenthal in Heffe is found in the form of ears of corn, and commonly called *kohrn-ahren*; and under the appearance of leaves and ftalks of fome graniferous vegetable.

Cinnabarine fhells are exceedingly **rare.**

I

I have in my **poffeffion fome** pfeudogalena of a blackifh yellow, united to millepores.

§ CLXXI. *Species of Phlogifticated organic Foffils.*

WOOD impregnated with petroleum frequently occurs. There is a trunk of a tree in the collection of the academy at Upfal, indurated with petroleum, black and fmooth, and yet eafily diftinguifhed to be of a beech. The Icelandic foffil wood alfo comes under this head, of which I have fpoken more fully in another place *.

Bones penetrated with afphaltus are fometimes found.

As is foffil wood likewife, whofe pores are filled with amber, and even with infects and other fmall animals; which this fubftance does not on**ly** penetrate, but even furrounds, as a fplendid monument covering their remains.

Turf and mould contain organic bodies, efpecially of vegetables reduced in the greateft part by putrefaction to duft; but which difplay figns of their original ftructure and character, **more** or lefs obfcure. The firft fcarce differs from the latter but in the greater decompofition and denfity of its mafs.

T 2   **VARIETIES.**

* Effays, v. iii. p. 239

V A R I E T I E S.

§ CLXXII.   *Ordinary Confusion of Varieties with Species.*

THAT many varieties have been obſerved in ſpecies properly determined is the more evident, as they have, for the moſt part, been conſidered as different ſpecies.   A miſtake to which the practice of the mineralogiſts in determining ſpecific differences from external marks undoubtedly gave riſe.

§ CLXXIII.   *Criteria of Varieties to be taken from external Appearances.*

IN the foregoing, we have ſhewn that ſpecific marks were to be taken from the particular compoſition ; but although ſuperficial criteria do not affect the intimate nature of theſe bodies, yet they are not by any means to be neglected ; they are well calculated to determine varieties, and are even uſeful, not only in leading often a ſkillful eye to proper diacritic experiments, but in throwing light upon the mode of production, and other intereſting circumſtances.

§ CLXXIV.

§ CLXXIV. *Illuſtration of external Marks.*

THE chief external marks are thoſe taken from the form of the outward ſurface ; the texture, in the appearance of its particles by a recent fracture ; the colour, hardneſs, and gravity.

§ CLXXV. *Amorphous Foſſils.*

FOSSILS that have no determined ſhape are denominated amorphous.

§. CLXXVI. *Chryſtalline Foſſils.*

BUT thoſe whoſe circumference is included within plain ſides meeting each other at various angles are called cryſtalline.

In the foſſil kingdom, we have five regular geometric figures, of plain, equal, and ſimilar ſides ; as the tetraedra, cubes, octaedra, dodecaedra, and icoſaedra ; beſides many others diſtinguiſhed by their priſmatic columns and pyramidal terminations. In what manner the great number of derivatives ariſe from a few primatives, and differing from each other at the firſt view, I have related elſewhere *.

Salts, indeed, on account of their ſolubility

T 3

in

---

* EſTays, vol. ii. p. i.

in water, more readily acquire a fubtilety and free-
dom of their particles, which, through the means
of attraction, is neceffary to form them into cryftalline concretions; but this property is not limited to them, as cryftalline foffils are found in almoft every genus of earths, metallic, and phlogiftic fubftances.

§ CLXXVII. *External Marks taken* **from the**
Texture of Foffils.

THE texture of foffils is not eafily determined
by the form of the particles; as when they are
intimately combined with each other they are
always mutilated by fractures; we may, however, diftinguifh many varieties. The moft fubtile fhapelefs molecules ufually called impalpable,
give rife to an equal texture: while others larger, and more difcernable produce a granous, fi-
lamentous, fcaly, and fpataceous compofition.

§ CLXXVIII. *From the Colour.*

COLOURS, efpecially the gradual fhades of
them, can fcarce be fo defcribed by language,
as to convey any clear idea. Hardly any other
method, therefore, than that of comparifon can
be ufed by always referring to thofe colours fufficiently underftood.

§ CLXXIX.

§ CLXXIX. *Phyfical Marks.*

PHYSICAL marks alfo, as hardnefs and gravity, are to be employed for afcertaining varieties,
whenever they are found to throw any light.

§ CLXXX. *Varieties of organic Foffils.*

THE varieties of organic foffils are to be determined from the fpecies of vegetables or animals,
which ferve as guides to our judgement. And
all living bodies being defined by their external
appearance, the fame rule may be obferved in
this as in the other claffes.

§ CLXXXI. *Epilogue.*

A SYSTEM of foffils, arranged according to the
foregoing method, I think is to be recommended for its variety, order, and utility; for the
number of fpecies and varieties, the manyfold
combinations of principles, the feries of agreement and difcrepancy, the harmony and oppofi-
tion of internal and external characters, and many
other important reafons: And I hope it will be
found to anfwer better, not only on account of
its extenfive view, but alfo becaufe the riches
and phenomena of the organic kingdoms are in
it more properly difplayed than in any other.

LATTER PART.

Of giving Names to Fossils.

§ CLXXXII. *The Utility of Names properly adapted in Mineralogy.*

If foffils are rightly and juftly arranged and
denominated, agreeably to the nature of things,
we find a harmony in them not lefs grateful
than advantageous.

§ CLXXXIII. *Hiftory of Names in Natural Philofophy.*

The fciences cultivated during the early a-
ges, as chemiftry, and all thofe depending on it,
had unhappily adopted certain fchemes and
modes of fpeech, of which the greater part were
not only puerile and abfurd, but often altogether falfe, and leading to erroneous conclufions.
Many circumftances contributed to the fupport
of this mummery. At firft, in thofe days of
darkeft ignorance, names were required to de-
fcribe new difcoveries and phenomena, adapted
to the unfkilfulnefs of their authors. By degrees the knowledge of natural bodies, as well
as of artificial, being extended, the profeffors of
chemiftry began to entertain fuch lofty ideas of
their

their fkill, that they did not hefitate to promife
themfelves the miracles of an univerfal medi-
cine, and the making of gold. Hence arofe
the ridiculous ftruggle betwixt the immoderate
boaftings, through which they were endeavour-
ing to difpofe advantageoufly of their difcoveries, and the moft folicitous attention with
which they wifhed to keep them concealed.
What the names they employed could be,
when depending on the moft abfurd theories,
the flighteft appearances, and moft abftrufe me-
taphors, we are at no lofs to apprehend. To
thefe were added afterwards others produced by
any fortuitous flight occurrence ; and we per-
ceive in fome meafure a language peculiar to
the early operations of chemiftry.

§ CLXXXIV. *Of reforming the Names of Foffils.*

THE inftitution of academies of fcience gave
rife to the gradual introduction of a founder
theory, founded upon more accurate experiment, which tended confiderably to limit the
barbarous and myftical affectation of fecrets; and
occafioned a more rational denomination of new
difcoveries, though as yet not built upon general principles. Befides, the rude and indigefted
mafs of antiquity was ftill preferved for the
greateft part, and chiefly for the following reafons. From the reformation of names and
phrafes

phrafes, it was apprehended that the whole
fcience would be involved in great confufion,
and that their number would create confiderable difficulties ; and it was likewife alledged,
that the moft ancient writings would, by this
means, be rendered unintelligible, and all the
fcience they contained condemned to oblivion.
But fuch evils, at leaft not all of them, feem not
to be a neceffary confequence. The oldeft
writings, efpecially thofe on alchemy, are almoft
all of them incomprehenfible: Whatever therefore will anfwer to probable conjecture, or will
admit of a certain and determinate explication,
might be more eafily underftood, if tranfpofed
according to the nature of the fubject,—and the
fenfe of this or that denomination being once
extracted, it might be preferved in a book ap-
propriated to the purpofe. As to what relates
to the dread of the introduction of new names,
it would undoubtedly be well grounded were
not all writers to fuffer them to be regulated in
the fame manner. In this cafe the new names
adapted to the nature of things would readily
infinuate themfelves, and be univerfally received.

Surely, it is highly improper that the nobleft
fcience, which conftitutes, as it were, the very
effence of natural philofophy, fhould deliver
truths of the greateft importance in the moft
abfurd of all languages. Every country in Europe

rope has thought the cultivation and perfection of its peculiar language an object highly worthy of attention; and shall the sciences alone be distinguished for rudenefs and barbarity of stile, while they are daily requiring new names to express new discoveries constructed upon rational principles; and which, if they are not all wisely and methodically ordered, would sometimes by their number occasion the destruction of those very discoveries they were intended to preserve. In botany, such a reformation has long taken place; and what is there that should prevent so salutary a plan from being extended to the other sciences?

But notwithstanding the obvious necessity of reform, as well as of some fixed standard, according to which all the new names should be regulated, there are still many difficulties that oppose their free introduction into the republic of letters. From the very nature of the proposal it is exposed to the influence of particular opinions; and every one, partial to his own, and chusing different data, it will be impossible in the beginning at least, to unite, in one common consent, sentiments so adverse and contradictory. We are not however to despair; for, if the voices of all do not combine, perhaps the greater number will, to stifle the clamour of persisting cavillers. Every real friend to chemistry, therefore, should wish for a happy issue to

the

the plan of Monf. Morveau, to be attempted in the new Encyclopædia. In the mean time, it may be permitted me to offer a few curfory remarks, which I think are relating particularly to mineralogy, and fubmit them to the judgement of the public. The end of the whole fyftem is doubtlefs to exprefs with truth, perfpicuity, precifion, and brevity, every thing of which an idea can be conveyed by words. New names, therefore, become neceffary to new things ; and to render thefe the moft convenient is the chief aim and object of this undertaking.

§ CLXXXV. *Names that are evidently abfurd, and ought to be expunged.*

I AM of opinion, that all abfurd names, and fuch as betray oftentatious vanity, are to be entirely fet afide. Of thefe we have examples in the fal mirabile Glauberi, fal fecretum Glauberi, fal polychreftum Glaferi, arcanum corralinum, arcanum duplicatum, fal de duobus, and feveral others.

§ CLXXXVI. *And falfe Names likewife.*

IN like manner, names that are falfe ought to be removed. Of this defcription are the following, fuggefting ideas that are erroneous :

Oleum

Oleum vitrioli	⎫	Concentrated vitriolic acid.
Spiritus vitrioli		Diluted vitriolic acid. Spirit indicates properly an inflammable liquor miscible with water.
Oleum tartari	For	Vegetable alkali dissolved by deliquescence.
Sal tartari		Alkali of tartar.
Terra foliata tartari		Acetous acid saturated with the vegetable alkali.
Butyrum antimonii		Muriatic acid saturated with antimony.
Semi-metallum	⎭	Fragile metal.

§ CLXXXVII. *What then are the names to be a-dopted?*

THOSE names which indicate some essential property or composition are of all others the best.

§ CLXXXVIII. *What are the Names to be tolerated?*

THOSE which admit a more extensive signification may be suffered, if others evidently better cannot be substituted. And these indeed

are

are true names; for although, from the power
of the words, they will apply to many fubftances, nothing prevents them from being και ιδιχην
applied to the one or the other. In this way
acidum aerium was ufed in the year 1772, for
aer fixum; which is not abfolutely advancing a
falfehood, as it poffeffes a proper acid, and in
an aerial form; but it is objectionable, becaufe
thefe qualities are difcoverable in other fubftances. Let therefore fome other denomination be
fubftituted more exact and determinate, as, gas,
or acidum mephiticum, or elfe there will be no
end to the various changes. But if it be impoffible to find one more accurate, it will be attended but with little inconvenience, to apply it to
that fubftance which we know for certain to be
the acidum aerium of the antients.

§ CLXXXIX. *Names fignifying lefs than the Thing
defined ought to be abolifhed.*

WHATEVER names exprefs too limited a fenfe
fhould certainly be expunged, if a choice can be
made among thofe that are fynonimous, efpecially thofe recommended by long time; as they
convey falfe and inadequate ideas. Thus *mineral* indicates properly an ore; but in the vulgar fenfe it fignifies every inorganic body found
in the bofom of the earth; although this idea is
more accurately expreffed by the word *foffil.*

In

In like manner, *oryctologia* implies a more exact denomination of the science of fossils than *mineralogia*. *Petrefactum* or *petrificatum*, falls nearly under the same criticism. But as here we have no better synonimous word to substitute, we must be contented with such as custom has established. Words, like coin, owe their currency to prescription.

§ cxc. *How we are to proceed without proper emphatic Names.*

As it is not easy to apply names exactly expressive of the thing defined, we are to adopt such as having no determinate meaning may have their sense ascertained by definition.

§ cxci. *Names derived from the Authors of new Discoveries.*

Among botanists and anatomists the memory of discoverers is perpetuated in particular denominations; it may, therefore, be a question, whether among chemists, where the reward of new facts is attended with greater inconvenience, it would be proper in the same manner to testify a grateful sense of obligation? To me, indeed, it seems to be practicable, and without any impropriety; but as it often happens, that the same discovery has been made by different individuals

individuals at the fame time, it might, upon the
whole, be better to truft the fame of all, to the
impartial records of the hiftoric page. This exception, however, need not extend to names of
little importance in chemiftry.

§ cxcii. *By what Means are the Claffes of Foffils to be defined ?*

Each clafs of foffils fhould, if poffible, be defined by one fingle word. Such as,—Salts,
Earths, Metals, and Phlogiftica. True, indeed,
the laft is an adjective ; but on this account folely it is not to be rejected, as we fhall prefently
fhew : Nor, indeed, have we reafon to apprehend ambiguity from the ufe of it, as the con-
text will always determine whenever **it** refers
to foffils. If any one fhould think the word
bitumina preferable, I can have no objections ;
although it may appear extraordinary to many
to confider diamonds under this definition.

For want of a more proper appellation, I diftinguifh foffils mixed mechanically under the
name of Petræ. My reafons for this diftinction
I have given already in § 166. Thofe, however,
that form the fubject of the other appendix, as
organic foffils, can fcarce be defined under one
title, and we muft therefore either employ two,
or call them in general Petrefactions, § 189.

§ cxciii.

§ CXCIII. *Denomination of Genera.*

EACH genus fhould be expreffed in one word, for the fake of brevity and convenience.

Among the falts there are, ftrictly fpeaking, but two genera ; the acid and the alkali. And we fhall fee by and bye the great advantage this produces, that the combinations of every acid conftitute proper genera. An acid may be confidered fubftantively without the neceffity of having the word Salt prefixed to it, as every acid is a falt.

In the fecond clafs we have found five genera. One of which, but lately difcovered, has, on account of its fpecific gravity, obtained the name of Terra Ponderofa. But in order to render it more concife and convenient, the firft word might be eafily omitted, though always underftood, and the laft employed alone as a fubftantive ; or we would, with Monf. de Morveau, adopt Barites from βαρυς with great advantage. The remaining earths are all expreffed with fubftantive names; but for the fake of perfpicuity, I would yet recommend fome alteration in them : As for example, Calx, Magnefia, Argilla, and Silex, are defcriptive of foffils, fuch as they occur on the furface of the earth, blended more or lefs with heterogeneous matter ; and therefore the words Calcareum, Magnefium,

U Argillaceum.

Argillaceum and Siliceum, might be properly u-
fed to fignify thefe fubftances pure and unmix-
ed.

The names of the fixteen metals are all fubftantives, and except one, are of the neuter
gender. The *ὑδραργυρος* of the Greeks was tranflated into Latin by Pliny *hydrargyrum*, and why
may not the *platina* of the Spaniards be adopted
into the fame language, with a neutral termina-
tion? According to this propofal, we fhall have the
following generic names, *aurum, platinum, argen-
tum, hydrargyrum, plumbum, cuprum, ferrum, ftan-
neum, vifmutum, niccolum, arfenicum, cobaltum, zin-
cum, antimonium, magnefium,* and *fiderum,* if this
laft differs at all from iron. Each of them are to
indicate the metal in its complete ftate. Dephlogifticated metals, commonly called calcined, or metallic calces, refemble indeed, in
fome meafure, burned chalk, from their attraction of the aerial acid, from their becoming cauftic with the volatile alkali, their fufceptibility of
pulverifation, and other properties.

Of phlogiftic bodies, the generic names are fo
well conftructed that we have no remarks to offer upon them : *Adamas, fulphur, petroleum,* and
faccinum, are received with propriety.

The four genera of *petræ* I define by the following names. The firft, abounding in faline
matter, I call *falfamentum* ; the fecond, loaded
with earthy matter, appears to me to be properly

ly *faxum ;* the third, containing metals in their matrices, I denominate *minera ;* and the fourth, from the mixture of petroleum, or other phlogiftic bodies more plentifully found in it, takes the name of *bitumen;* or, if this name be given to a clafs, *picarium* may be fubftituted.

Of the organic foffils, that which is penetrated with any falt may be called *falitura ;* with earthy particles, *lapidofum ;* with metallic, *metalliferum;* and with phlogiftic, *pollinctum.* Should names more proper than thefe occur to any perfon I fhall have no objection to withdraw them.

§ CXCIV. *Of applying Names to the fimpler Foffils, and efpecially to the Salts.*

ALL bodies, whofe proximate principles have never yet been afcertained by art, require fimpler names; the primitives efpecially fhould be exprefled by one word; and thofe of a known compofition fhould be defined by derivatives having a reference to their principles; if not **of** one or two words, confifting at the moft of three. To denote each body by a peculiar fimple name **would** be productive **of** great inconvenience, and **be** an ufelefs burden to the memory. It might however be of confiderable advantage to the fyftem of nomenclature, in the clafs of falts, if every one of the fimple falts could be indicated by a fingle word. Would it

not

not therefore be admiffible, by fuppofing the
acid to conftruct the names as fubftantives? As
for example, *vitriolicum, nitrofum, muriaticum,
regalinum, fluoratum, arfenicale, boracinum, fac-
charinum, oxyalinum,* (inherent in the acid of the
wood forrel) *tartarum, benzoinum, citrinum, fuc-
cineum, galacticum, formicale, febaceum, phofpho-
reum,* and *aereum.* Phlogifticated vitriolic acid
might be named *fulphureum,* and phlogifticated
nitrous acid *nitreum.* In like manner, in the
genus of alkalies, the vegetable will be *potaffi-
num;* the mineral *natrum,* a name by which it
has fometime been already known; and the **vo**-
latile will be *ammoniacum.* The great advantage of this fimplicity, as we fhall fee prefently,
will be obvious in giving names to compounded
fubftances; which, if they confift of more than
two or three words, will give rife to a diffufe and
circuitous ftile, both in fpeaking and writing.
All names certainly proceeding from the definition of feveral words are by far the moft improper.

§ cxcv. *Names of Species demonftrated in the
Cafe of Salts.*

SPECIFIC differences, that can ferve as diftinct names, are ufed with confiderable advantage. Admitting what has been already propofed in the preceding paragraph, this very eafily

fly obtains in the clafs of falts, as to all the fpecies perfectly faturated. That earthy and metallic falts ought to be arranged under the head of their menftrua, we have feen in § 70.; but, with refpect to the perfect neutral falts, it is not fo clear, § 68. It feems indeed more convenient to refer them to the genera of their feveral bafes ; and in this way alfo I have proceeded. But we fhall have more agreement with the analogical falts, moft of which are properly affigned to the acid, if the neutral falts are fubjected to the fame arrangement. According to this method we fhall have names fufficiently apt by combining the acid with the adjective of the bafis. As for example,

Vitriolicum potaffinatum,	for	{ Tartarus vitriolatum.
Nitrofum natratum,	—	{ Nitrum cubicum.
Muriaticum ammoniacum,	—	{ Sal ammoniacus.
Acetum potaffinatum,	—	{ Terra foliata tartari.
Vitriolicum calcareatum,	—	Gypfum.
—————— magnefiatum,	—	{ Sal catharticus amarus.
————— argillatum,	—	Alumen.
Nitrofum barytatum,	—	{ Barytes nitratus.
————— argillatum,	—	{ Calcareum nitratum.
Muriaticum barytatum, &c.	—	{ Barytes muriaticus.

Metallic

Metallic double falts alfo may be treated in the fame manner ; as,

> Vitriolicum auratum, &c.
> Nitrofum argentatum, &c.
> Muriaticum plumbatum, &c.
> Arfenicale cobaltatum, &c.

and many others.

No one can object to thofe adjectives derived from the names of the metals, as Pliny ufes the word *ferratum ;* and it is according to this plan that they are here applied.

Analogical falts, containing an alkali, may be eafily arranged in the fame manner.

Thus,

Potaffinum	—	Argillatum, Silicatum, &c.
Ammoniacum	—	Argentatum. Cupratum, Zincatum, &c.

Double falts, in which either principle prevails can alfo be denominated in fuch a manner as to exprefs an imperfect faturation, § 127. For example,—Tartar, with an excefs of acid, can be defined by a combination of its generic name with the genitive of its bafe, as *tartareum potaffini ;* but, when perfectly faturated, may be called *tartareum potaffinatum.* In like manner we fhall have *oxalinum potaffini,* but, when exactly faturated, it will be *oxalinum potaffinatum ; vitriolicum natri,* and *vitriolicum natratum ; natrum boracini,* and *boracinum natratum ;* and fo on of others.

This

This method, however, is not applicable in o-
ther claffes, not even to the double fpecies. Saline earths, with fuch an excefs of earthy matter
as nearly to obliterate their faline character,
ought thus to be expreffed.

Barytes vitriolatus,   for Spatum ponderofum.
Calcareum fluoratum, —— Fluor mineralis.
Calcareum aeratum, —— Calcareum vulgare.

The character of the remaining foffils differs
more confiderably from the falts, and requires
auxiliary illuftration.

§ cxcvi. *Trivial Names of Salts.*

Fossils, containing three or more principles
appear capable of the cleareft definition by
means of the trivial names. The celebrated Linnæus firft made ufe of fuch, in his Species Plantarum of 1753, by which every fpecies
could be conveniently expreffed, without a repe-
tition of the fpecific differences. The language
of botany became thus remarkably eafy and intelligible ; and zooligifts and mineralogifts have
to thank the fame author for the happy introduction of them into their fciences.— But, although thefe names may be affumed from the
inventor, fome virtue, ancient appellation, property, or accidental circumftance refpecting the
fpecies; yet fhould they be generally limited to
one word, and very feldom indeed extend to
two. They may be confidered as furnames dif-

U 4   tinguifhing

tinguifhing the individuals contained in the fame genus.

The triple falts are, by means of thefe trivial names, denominated with great facility.   Of which we have the following examples :

Vitriolicum fallax	—	Epfom falt united to the volatile alkali; eafily producing an apparent inequality of attract.
―――― epilepticum,	—	Epileptic falt of Weifman.
Muriaticum anti-epilepticum,	—	Anti-epilepticum puerorum of Boerhaave.
―――― alembrot,	—	Sal alembrot.
―――― dulce,	—	Mercurius dulcis.
Galacticum Bartoleti,	—	Sugar of milk, firft defcribed by Bartoletus.
Tartarum Seignetti,	—	Sal polychreftum Seignetti.
―――― Lafonii,	—	Tartar joined to the fedative falt.
―――― folubile,	—	Tartar faturated with volatile alkali,— commonly called tartarus folubilis.
―――― Mynfichti *,	—	Tartarus emeticus.
―――― martiale	—	Globuli martiales.
Phofphorum microfmicum,	—	Sal microcofmicus.

Compound

* Effays, vol. i. p. 340.

Compound falts, produced by regalinum (a-qua regis) never become triple, at leaft not all of them. The nitrous acid feems to be neceffary for the purpofe of dephlogiftication only; and the muriatic generally exhibits the fame combinations as the regaline, by which, if the muriatic is not in fufficient quantity, a double falt is obtained, charged with the nitrous acid,

The fame obfervation is equally applicable to the quadruple falts.

Tartarum Fevri,	— {	Tartar united to borax,
Nitrofum Kunckelii,	—	Rubini Kunckelii.
————— fympathicum,	— {	Sal ammoniac with nitrated copper.

This falt exhibits cryftals, that affume a yellow colour **when** heated, but become blue in a moderate temperature. If a folution of them fufficiently diluted is ufed for writing, the letters will be found to difappear entirely, by the application of heat; and, if expofed to the vapour of cauftic volatile alkali, to change to a beautiful blue colour.

Thus, then, I have pointed out a method, as I apprehend, both eafy and fimple, by which all the known falts, about fifty in number, may be each denominated in one or at moft in two words.—According to the firft divifion, we have the genus only.—Of the fecond, the double falts completely faturated are indicated by the adjective of their bafe ending **in** *atus*. In the third, the imperfect falts are known by the genitive of their bafe.—The fourth contains the triple falts and thofe of feveral principles, which are expreffed by the trivial names; and as in them we neither find the adjective of the bafe *atus*, nor the genitive, it is not poffible that any ambiguity can arife.—The whole compofition of the triple falts could not be fignified in two words, unlefs the double falts were defined in one only; and if the fame brevity were expected of the quadruple, the triple muft have necef-farily been denominated by one. But it may be a queftion, whether it is more difficult to invent fuch a number of new and fimple names, or, if invented, whether they could poffibly be retained by the memory.

§ CXCVII. *Of the fpecific Names of Earths, Metals, and Phlogiftic Subftances.*

IF we confider every thing that has been faid in the foregoing fections on the fubject of the falts hitherto known and inveftigated, we fhall find, that we have in fome meafure laid the foundation of a general fyftem of mineralogy. With regard to the earths, and the following claffes, the denomination of the double and more compounded fpecies may be conveniently expreffed by the trivial names in two words. Thus, for example, under the genus magnefia, a fpecies occur, in the formula, *fmca*, compofed

of

of filiceous, calcareous, and argillaceous earth, with fome admixture of iron *, which in fyftematic authors is denominated afbeftos, and treated as a peculiar genus. To this, indeed, the trivial name of afbeftos may be properly applied, as it feems to be fo well underftood, that the youngeft mineralogift is in no danger of being mifled by it. The fame may be faid of fchoerl, granate, zeolite, and many others, that are diftinguifhed by names known to every body, and highly proper. In the compofition of earths, iron is by no means a neceffary ingredient, although it is generally found in them; and we therefore confider it as an alloy, or heterogeneous fubftance.

§ CXCVIII. *Conclufion.*

I CANNOT finifh my remarks on the denomination of foffils more to my own fatisfaction, than by pointing out what is yet wanting to the improvement of fcience. I would wifh that in the eftablifhing of new names, a preference fhould be given to the Latin language. This is, or at leaft was formerly the mother tongue of the learned; and being now not the living language of any nation, it is no longer liable to innovation or change. If therefore, the reform we propofe is made firft in Latin; it may be eafily

carried

---

* Differtation on the afbeftos.

carried into execution afterwards upon the fame model in the modern languages, as far as their peculiar genius and conftruction will admit.—In this manner, the language of chemiftry will become every where uniform and confiftent, and confiderable advantage will be derived not from the reading only of foreign publications, but the facility alfo with which they can be tranflated. I have feen an excellent effay of Monf. de Morveau on the reform of the French names*, and I am not a little flattered by the agreement I find between many of the alterations he propofes and thofe that I have offered on that fubject. From this, perhaps, we may venture to hope, that by making it an object of further attention on both fides, the differences yet fubfifting may be removed, to the great benefit of fcience; and to the permanent eftablifhing and advancement of which all our views fhould be directed.

---

* Diary of Monf. Rozier.

www.ingramcontent.com/pod-product-compliance
Lightning Source LLC
Chambersburg PA
CBHW021825190326
41518CB00007B/741